Practical Guide to Comparative Advertising

Dare to Compare

Practical Guide to Comparative Advertising
Dare to Compare

RUTH M. CORBIN
Chair, CorbinPartners Inc., Adjunct Professor Osgoode
Hall Law School, and Mediator, Corbin Estates Law

REBECCA N. BLEIBAUM
President/Chief, Sensory Intelligence of Dragonfly
SCI, Inc.

TOM JIRGAL
Partner, Loeb and Loeb LLP

DAVID MALLEN
Partner, Loeb and Loeb LLP

CHRISTINE A. VAN DONGEN
Formerly Head, Sensory and Consumer Preference,
Nestle Health Science R&D Center, Minneapolis

ACADEMIC PRESS
An imprint of Elsevier

Academic Press is an imprint of Elsevier
125 London Wall, London EC2Y 5AS, United Kingdom
525 B Street, Suite 1650, San Diego, CA 92101-4495, United States
50 Hampshire Street, 5th Floor, Cambridge, MA 02139, United States
The Boulevard, Langford Lane, Kidlington, Oxford OX5 1GB, United Kingdom

Notices
Knowledge and best practice in this field are constantly changing. As new research and experience
broaden our understanding, changes in research methods, professional practices, or medical treatment
may become necessary.

Practitioners and researchers must always rely on their own experience and knowledge in evaluating
and using any information, methods, compounds, or experiments described herein. In using such
information or methods they should be mindful of their own safety and the safety of others, including
parties for whom they have a professional responsibility.

To the fullest extent of the law, neither the Publisher nor the authors, contributors, or editors, assume
any liability for any injury and/or damage to persons or property as a matter of products liability,
negligence or otherwise, or from any use or operation of any methods, products, instructions, or ideas
contained in the material herein.

British Library Cataloguing-in-Publication Data
A catalogue record for this book is available from the British Library

Library of Congress Cataloging-in-Publication Data
A catalog record for this book is available from the Library of Congress

ISBN: 978-0-12-805471-0

For Information on all Academic Press publications
visit our website at https://www.elsevier.com/books-and-journals

  Working together
to grow libraries in
developing countries

www.elsevier.com • www.bookaid.org

ASTM INTERNATIONAL

Publisher: Andre Gerhard Wolff
Acquisition Editor: Nancy Maragioglio
Editorial Project Manager: Billie Jean Fernandez
Production Project Manager: Vijayaraj Purushothaman
Cover Designer: Victoria Pearson

Typeset by MPS Limited, Chennai, India

CONTENTS

PREFACE

Better. Faster. Stronger. Whiter. The one recommended by more doctors, more dentists, more hairdressers. Advertised comparisons like these let customers know why your company's products or services are the ones they can trust. Comparative advertising can sometimes feel like a thrill ride, combining the excitement of gaining new customers with the dangers of inciting litigation-minded competitors.

This is a handbook for corporate management teams—the advertisers, researchers, in-house counsel, and marketing executives who dare to lob a comparative advertisement that declares or implies their superiority over competitors. It is designed to be both practical and authoritative—a book that management professionals can keep handy on their shelf, for easy constant reference. Each chapter concludes with a checklist to remind readers not only of what steps to take but also of what risks to avoid.

I am indebted to my "dream-team" of collaborators, each of whom has brought distinct expertise and the war-stories of experience to this dynamic field of corporate communications.

Ruth M. Corbin
June, 2017

CHAPTER 1

Comparative Advertising—Look Before You Leap

Comparative advertising promotes an organization's products or services by reference to the products or services of others. By its very design, it encourages people to make comparisons. It can explicitly name a competitor or merely imply superiority or comparisons through its language, like "better-tasting," "#1 recommended," or "lowest price." Memorable comparative advertising campaigns pervade the popular culture. The Pepsi Challenge (*vis-à-vis* archrival Coca-Cola) has been bubbling along for about 40 years, revitalized every so often with new technology and promotional ideas.

A new-age cyber version of a similar competitive dare has appeared in the form of Microsoft's "Bing it On Challenge," inviting Internet users to compare the value of Bing versus Google search results.[a] Although there is no single measure of the success of Microsoft's initial campaign, the

---

[a] A video of the Bing Challenge in action in Britain is shown at https://www.youtube.com/watch?v = AgeKTTP9_KM, accessed June 19, 2016.

Ruth M. Corbin, (Editor): Practical Guide to Comparative Advertising
Comparative Advertising—Look Before You Leap, (Dr. Ruth M. Corbin), Principal author.
DOI: https://doi.org/10.1016/B978-0-12-805471-0.00001-7. © 2019 Elsevier Inc. All rights reserved.

campaign slogan "Bing it on" has powerfully endured, headlining reputable reports of Bing's growth at the expense of Google[b]

Bing It On
A battle of two search engines side by side.
Go through the test, and vote to see which one's better

Commercials featuring the cool Mac dude versus the stodgy PC guy attracted such audience affection, that more and more advertisement scenarios were produced. How many ads in the series in all? 66!

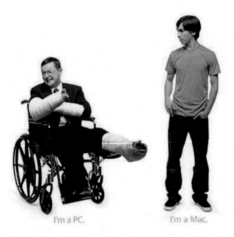

I'm a PC.   I'm a Mac.

In 2011, the National Australia Bank (NAB) won Cannes' prestigious Grand Prix as that year's "best campaign in the world" for its series of advertisements announcing a breakup between it and the four largest banks in the country. The tongue-in-cheek campaign featured "breakup" letters, conversations, and messages that people might use to escape a bad relationship. The media success motivated NAB to launch a website dedicated to the campaign, at http://breakup.nab.com.au/, bearing the headline "We're not popular with the other banks anymore. We must be

---

[b] See, for example, USA Today's posted article in April 2015, entitled "Bing it on. Google loses 20%," available at http://americasmarkets.usatoday.com/2015/04/16/google-gets-dinged-by-bing/. Search Engine usage is tracked by comScore, an independent media measurement company, as reported at http://www.comscore.com/Insights/Market-Rankings/comScore-Releases-November-2015-US-Desktop-Search-Engine-Rankings, accessed June 19, 2016. The graphic below text at www.bingiton.com, accessed August 7, 2016.

doing something right." News in 2012 conveyed that NAB had attracted its millionth new customer since the campaign commenced.[1]

Does comparative advertising "work?" It apparently does work, at least sometimes, according to the examples above. The fun, the glamour, or the bravado of comparative advertising offer further temptation to consider it for one's next big advertisement campaign. But the decision to launch a comparative advertisement is a commitment to something more than just communicating a message. It is a commitment to a style of communicating and to taking on risk. The decision has so many possible consequences for an organization that it would be unwise to proceed on the path of comparative advertising without the endorsement of the chief executive. A chief executive sets the tone and culture of an organization, more pervasively than some may realize.[2] Compatibility with the organization's tone and culture reflects ultimately on the clarity of its entire brand positioning (Box 1.1).

BOX 1.1 Case in Point

Worldwide diversified consumer products company SC Johnson thrives on advertising its product superiority. It has publicly declared its commitment to developing products and services that are recognized by consumers as "significantly superior" to major competition.[c] Encouraging comparisons between its products and those of others is inherent to its advertising strategy.

(Continued)

---

[c] "This we Believe," SC Johnson's statement of corporate values, published at http://www.scjohnson.com/Libraries/Download_Documents/TWB-English.sflb.ashx, last visited November 30, 2017.

BOX 1.1 (Continued)

But how to make comparisons without getting challenged? It's a matter of balance, said one SC Johnson spokesperson. The company chooses where to take risks. It invests in testing up front. Its comparative advertisements emphasize product benefits and appeal to consumers' common sense. Interestingly, SC Johnson commercials seldom if ever do side-by-side comparisons or call out competitors explicitly. Its advertising culture encourages respectfulness, "winning in the right way."[d]

Launching a comparative advertisement campaign within a large corporation requires an ordered process of idea generation, claim development, advertising design, legal clearance, and disaster-checking before going live. Many departments at SC Johnson are involved in a team approach, including R&D, marketing management, legal, and marketing research. The marketing department usually leads the team. They are the ones who bring to bear the perspective of the customer. They help to define "success" for the corporation. What integrates the team of professionals from different departments is a collective goal of success.

For testing standards, SC Johnson looks either to ASTM or to sector-specific research standards, like the ones associated with ISO. The ultimate criterion, even with scientific support, is what consumers understand about the claim itself.

Its diligence in supporting advertising claims extends to holding competitors to account for their own claims of product superiority. Among its competitive battles, SC Johnson disputed in court a comparative advertisement for resealable plastic bags. An advertisement campaign by competitor Clorox compared its GLAD Bag to SC Johnson's ZIPLOC bag, by showing each bag holding an animated goldfish in water. The bags were turned upside down. In the commercials, Clorox's bag stayed tightly closed, while the ZIPLOC bag was shown to purportedly leak water at a troubling rate, much to the panic of the goldfish inside. An expert hired by SC Johnson gave testimony in court that the commercial exaggerated whatever test result Clorox might have had, to an extent that would surely mislead consumers. He tested 100 ZIPLOC bags for leakage and found that 37% did not leak at all, and of the remainder, the vast majority leaked at a rate between 2 and 20 times slower than that depicted in the Clorox commercial. The court, in that case, concluded that the Clorox commercial was both false and misleading, and that it had to be taken

(Continued)

---

[d] That is how the company communicated its culture to prospective employees, available at https://www.linkedin.com/jobs/view/director%2C-claims-%26-sensory-at-sc-johnson-475317309/, last visited November 30, 2017.

BOX 1.1 (Continued)

off the air. More details about that case and a follow-up to it are available at https://www.leagle.com/decision/2001473241f3d2321447.[e]

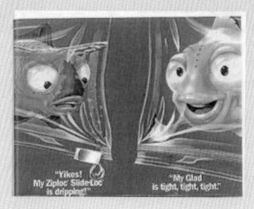

IT'S A WAR OUT THERE

Comparative advertising conveys a boldness of style. Consider how frequently it has been described with battleground analogies. "Let the wiener wars begin," said Judge Morton Denlow as he opened the 2011 US trial about the comparative advertising of Ball Park Franks and Oscar Meyer Wieners. CBC News ran the story in Canada,[3] observing that the "the battle pits...Sara Lee Corp, which makes Ball Park Franks, against Kraft Foods Inc., which makes Oscar Mayer," and further described how Sara Lee had "fired the first volley."

Consider a further illustration, in the following effusive paragraph from an article[f] by advertising expert, author, and film-maker Herschell Gordon Lewis:

CHOOSE YOUR WEAPONS: THE FOUR FACES OF COMPARATIVES

Comparative advertising can be a formidable weapon in the hands of a trained marksman. But if you don't involve the reader in your claims of superiority, you may be shooting yourself in the foot.

(Continued)

---

[e] Last visited October 9, 2017.
[f] *Direct Marketing Magazine,* May 1990.

> **(Continued)**
> *In any battle, the most potent weapon is also the most dangerous weapon. It can kill you instead of your target, especially if you let half-trained troops load, aim and fire it.*
> *That's absolutely true of comparative advertising. Comparisons are an exquisite rapier. A skilled copy-swordsman can run the competition through with a deft thrust; an inept word-pusher, swinging comparatives with saber-like wildness, can kill his copy with inept swordplay.*

Headlines such as "Attack ads," "cola wars," "battle of the brands," "marketing warfare," "battle of the burgers" for books and news articles depict comparative advertising as an act of "confronting the enemy." It's a style that needs to sit well with your company, as a matter of strategic positioning and corporate culture.

FACTORS TO CONSIDER

The choice to be a "comparative advertiser" has implications much beyond the marketing department. Customer perception, employee morale, industry relationships, legal position, and financial risk/reward are all implicated in the choice. Each is considered in the following sections:

Customers Notice

Consumers are seldom passive accepters of "facts." Encountering a comparative advertisement, they form views about its usefulness and truthfulness, and about the company behind it. They filter what they hear or see through the lens of their own experience and values.

Diverse published research[g] on how customers view comparative advertising yields the following general purpose conclusions, subject of course to case-by-case exceptions:

- Customers appreciate receiving factual comparative information about product features that help them find the product that is personally right for them. They like to feel "smarter," and more in control of making good decisions.

[g] For a comprehensive assembly of pre-1997 published research, see Ref. 4.

- Comparative advertisements are more effective than noncomparative advertisements in attracting attention, facilitating recall, increasing the likeability of the advertised brand, and increasing purchase intentions.
- Based on the surveys of how much consumers "like" different advertisements, reported likeability is lower for comparative advertisements. The believability of the source is also generally rated lower. Advertisers should note, however, that likeability of advertisements, as reported in surveys, is not a consistent predictor of what consumers actually buy.
- Advertisements that go beyond factual comparisons to comment negatively on a competitive brand or to devalue a competitive brand by innuendo, frequently meet with customer disapproval.
- In extreme cases, a comparative advertisement may backfire by engendering public sympathy for a maligned brand.

The trade-off decision left to the advertiser, then, is whether to seek the reward of more business from customers, at the expense of being possibly viewed as a negative or mean-spirited advertiser. The risk of the latter can be mitigated by the use of demonstrably factual information, humor, and respectful tone.

Employees Love a Winner

Employee engagement is enhanced by corporate culture and corporate values with which employees can identify.[h] These seem particularly important today to younger professionals in the workforce. Although employees are therefore likely to enjoy advertisements which convey their company's competitive superiority, advertisements that are embarrassing, disrespectful, or mean-spirited will insert a negative element into the company—employee relationship. Previewing comparative advertisements to employees, prior to their launch, with an explanation of the advertising objectives, will help to engage them in informed, even enthusiastic, dialog. It also prepares employees for comments by their friends and acquaintances.

---

[h] For one of many well-researched books on the link between employee engagement and business strategy, see Ref. 5.

Legal Standards Differ Around the World

It is no coincidence that "comparative advertising" has been explicitly defined in the laws, directives, or regulations of several countries. That is warning in itself to corporations that complications lie ahead.

The United States and Canada are among the most constructively supportive countries, giving priority to "consumer benefit" of comparative advertising.[i] Countries of the European Union, while also embracing the benefits of comparative advertising, include an explicit directive against the discrediting of competitors' trademarks.[j] Russian law does not mention "comparative advertising," yet has regulations against fraudulent publicity that discourage negative comparisons of any type. Journals and websites present articles from several other countries distinguishing the nuances of allowable trademark use in comparative advertising in their home jurisdictions. Cataloging them is beyond the scope of this book. The relevance to the present topic is that a multinational company launching a comparative advertisement in different countries must systematically check whether its use of a competitor's trademark will stay within each country's laws.

Multicountry campaigns are vulnerable to an equally important issue of cultural acceptance. Comparative advertising is less likely to be effective in cultures that value harmony and discourage confrontation or those with "collective" rather than "individualist" orientations to society.[k] Offense to cultural attitudes means damage to the advertiser's trademark

---

[i] See FTC's "Statement of Policy Regarding Comparative Advertising," 1979, at Paragraph c; Regarding US/Canada similarities, see Pritchard and Corbin. "Avoiding Landmines. How to Support Comparative Advertising Claims," presentation to a conference sponsored by Advertising Standards Canada, February 27, 2012. Content available online in slide form at http://corbinpartners.com/wp-content/uploads/2012/12/ASC-Presentation-Avoiding-Landmines-How-to-Support-Comparative-Advertising-Claims.pdf as on June 2016.

[j] The European Union's "Directive 2006/114/EC concerning misleading and comparative advertising," at the time of this writing, United Kingdom is a member of the European Union, but its regulations are more explicit in cautioning against abuse of competitors' marks.

[k] Korea is one such culture, as described by marketing text authors Wayne Hoyer and Deborah MacInnes in *Consumer Behavior* (2008), Manson, Ohio: Cengage Learning, at page 135. Author Meghna Singh of the University of Delhi presents a broad-based analysis of comparative advertising effectiveness across different countries and cultures in "Comparative Advertising Effectiveness with Legal and Cross-culture Framework" (2014), *International Journal for Research in Management and Pharmacy*, Vol. 3, Issue 3.

value. Protecting that value goes beyond merely ensuring that the advertiser stays on the right side of the law.

In summary, adapting comparative advertising to other countries is more than just language translation; it requires an investment in customization of laws and cultures of each country.

Industry Relationships Are Affected by Advertising Wars

The benefits and personal satisfaction that executives in some industries discover, in cooperating with competitors on matters of legitimate common interest, have given rise to the concept of "co-opetition".[6] Benefits could include cooperative advocacy on industry regulations (such as the occasion when Canadian telecom companies cooperated in their policy advocacy regarding foreign entrants), mutual business support (such as hotels referring their overflow customers to other hotels in their area), sharing of industry trend information (such as adaptation of environment-friendly substitutes for chemical compounds), or collaborating to increase the size of the "market pie," so that each competitor's market share is thereby bigger in absolute size. Aggressive comparative advertising may interfere with the ability to strike up beneficial cooperative relationships. It is difficult to sit at a boardroom table sipping coffee with a competitor who has described your products as inferior or overpriced.

Budgets Demand More Than Standard Advertising Costs

Comparative advertising requires an ancillary budget for all of the following reasons:

- A competitive comparison should be made only with support in hand. An investment in new data collection may be necessary.
- A comparative advertisement is a time-limited resource. As market circumstances change, the facts in the advertisement may no longer be true. Market monitoring should be undertaken to ensure that the comparison remains true. The life span of a comparative advertisement is generally thought to be about a year.
- Be prepared for a fight even if your legal position is sound. The risk of a complaint by a competitor is higher if the competitor is named. Legal battles can run into the hundreds of thousands of dollars in cost.

Although advertising budgets would not normally account for maximum financial penalties of losing a legal battle, one should at least be aware of the risks at hand. In 2011, Canada's Competition Bureau levied a fine of 10 million dollars against Bell Canada for misleading advertising. The object of its displeasure was Bell's competitive pricing advertisements. The Bureau imposed on Bell the further indignity of another $100,000 to pay the costs of the investigation.

MOST PROMISING OCCASIONS FOR COMPARATIVE ADVERTISING

If your company is ready in principle for comparative advertising, it is useful to recognize the occasions where it is most likely to be all-round effective. Published experience suggests that the following factors enhance the likelihood of a successful comparative campaign:

- Comparative advertisements are more likely to be appreciated and recalled when they target a known area of customer dissatisfaction, such as high prices. For example, a Canadian newspaper recently carried the announcement that Pfizer's Viagra drug will now be priced to be the same as its generic substitutes.
- Lesser known players gain from positioning themselves against a well-known brand, thereby giving instant boost to their profile.[7]
- "Objective" brand attributes and messages are likely to be more effective than "subjective" or "emotional" attributes and messages.[7]
- Where advertising content goes to something other than objective attributes or facts, believability and humor have been shown to sit well with the consumers.

The main message for advertisers is that comparative advertising is an invitation to battle that is not for the faint of heart or the short of cash.

THE TRAJECTORY OF A COMPARATIVE ADVERTISING CAMPAIGN

The chapters in this book offer an educational tour through the steps and issues for launching a comparative advertisement campaign:

- First, be aware of legal and regulatory restrictions for what can be claimed in a comparative advertisement. Chapter 2, Governing Laws, Network Standards, and Industry Self-Regulation, provides a brief education on that topic.
- Next, consider what claim of superiority over a competitor you want to make. The claim will likely complement the position you seek for

your product or company overall. Chapter 3, What's the Name of the Claim, offers a catalog of claims from which to choose.

- Claims must be explicitly supported by evidence. Chapter 4, Foundations of Test Design, describes the criteria that your evidence must address, to withstand scrutiny before a regulator or judge.

- Comparative claims inevitably involve quantitative words like "more," "majority," or "most preferred." Proving your claim will come down to statistics of some kind. How big does your statistic have to be to qualify? The answer is less obvious than you think. See Chapter 5: Statistical Support—How Much Is Enough?.

- With the preparatory steps above, a company is ready to add the magic of a brilliantly designed advertisement. There is no chapter in this book on summoning up the magic. But once designed, a comparative advertisement needs to be fine-tuned. It may need, for example, explanatory notes or disclaimers to avoid misleading the audience. It may need occasional tweaking to avoid obsolescence, particularly if your competitor makes changes in the very product being compared in your advertisement. Chapter 6, Know Your Limits: Claims Have Boundaries, reminds you about the fine-tuning issues.

- Anticipate the unexpected. Even when everything looks fine to the corporate team, a final disaster check is in order before launch. Chapter 7, An Ounce of Prevention: Troubleshoot Your Claim Before Launch, advises on those final steps to minimize the chance of a market fiasco once the advertisement is launched.

- Comparative advertisements provoke competitors. Once the advertisement is launched, marketers and their legal advisors should reserve time and money for addressing a possible legal challenge by the targeted competitor. See Chapter 8: Into the Fray: Playing Defense, on how to prepare.

- What if your company is the targeted competitor, rather than the protagonist, in a comparative advertising campaign? Chapter 9, Into the Fray: Playing Offense, provides advice on reacting to such a campaign.

- Corporations doing business in more than one country will find useful the international perspectives added in Chapter 10, *Vive la Difference—* Adapting Comparative Advertising to Different Countries.

- Chapter 11, Advertising Claims in Social Media, treats the special, although increasingly common, context of social media advertising. What happens when consumers carry your message through their own unpredictable conversations?

- A practical checklist of steps appears in Chapter 12, Summary and Handy Checklist. Once you have benefitted from a review of the step–by–step chapter topics, Chapter 13, Twenty–First Century Resources, may be the only reference you will need.
- A curated bibliography of resources published in the 21st century, from articles to videos to media reports appears as Chapter 13, Twenty–First Century Resources. The set of resources offers a starting point for further research, and a compendium for experts to consult should they be called upon for adjudicating opinions.

"The Granddaddy of Comparative Advertising"
The modest beginnings of comparative advertising are frequently credited to J. Stirling Getchell, hired by Chrysler in 1932 to promote its 3-year old Plymouth brand. Plymouth was Chrysler's low-end car that competed with models by industry leaders Chevrolet and Ford.

Getchell and his partners photographed a line-up of all three cars, under the headline "Look at All Three—and May the Best Car Win." The text boasted "Plymouth Sets the Pace for All Three." Only Plymouth was ever named in the accompanying text, competitor brands were shown but not mentioned by name. Despite the low-key comparison, the innovative advertisement engaged prospective buyers. According to historical accounts, Plymouth's sales rose by 218% in the 3 months after the advertisement campaign; its market share also began a steady upward rise. The success of the campaign was confirmed in a shout-out "Thank you" by the President of Chrysler in an advertisement appearing in *Washington Post* on April 24, 1932, at page 8.

The 1932 Plymouth campaign holds iconic status in the history of advertising. Author Julian Watkins devoted a chapter to the "Look at all Three" advertisements in his book on "The One Hundred Greatest Advertisements."[1] "Almost everybody in advertising will admit that the late J. Stirling Getchell's advertisement for Plymouth 'Look at all Three', is one of the all-time greats.," he wrote. A journal article by the president of agency Cunningham & Walsh Inc., Anthony Chevins, referred to the advertisement as "what may be the grand-daddy of all specific comparative ads".[8]

(Continued)

---

[1] Mineola, NY: Dover Publications, 1959.

(Continued)

REFERENCES

1. Murdoch S. National Australia Bank set to sign up its one millionth customer since the 'break-up' campaign. *The Australian* July 24, 2010.
2. Berson Y, Oreg S, Dvir T. CEO values, organizational culture, and firm outcomes. *J Org Behav* 2008;**29**:615−33.
3. Associated Press. *Communicated by CBC News/Business.* Available at: <http://www.cbc.ca/news/business/story/2011/08/15/kraft-sara-lee-hot-dog-wars.html>; 2011 [accessed 27.11.12].

4. Grewal D, Kavonoor S, Fern EF, Costley C, Barnes J. Comparative versus non-comparative advertising: a meta-analysis. *J Mark* 1997;**61**:1—15.
5. Cook S. *The essential guide to employee engagement: better business performance through staff satisfaction.* London: Kogan Page; 2008.
6. Brandenburger A, Nalebuff B. *Co-opetition: a revolution mindset that combines competition and cooperation: the game theory strategy that's changing the game of business.* New York: Doubleday; 1996.
7. Barry TE. Comparative advertising: what have we learned in two decades? *J Advert Res* 1993;**33**(2):19—29.
8. Chevins A. A case for comparative advertising. *J Advert* 1975;**4**:31—6.

CHAPTER 2

Governing Laws, Network Standards, and Industry Self-Regulation

OVERVIEW

Chapter 1, Comparative Advertising—Look Before You Leap, explained the factors for deciding when comparative advertising makes good business sense. Once a company sets out on a path of comparative advertising, it needs to give early attention to its obligations under the law. This chapter outlines the relevant principles, laws, and regulations, and also describes the process of self-regulation that gives rise to many of the standards that help govern responsible advertising.

Comparative advertising, as with all marketing practice, is subject to laws designed to protect consumer welfare and to safeguard companies from unfair competition. Most countries permit comparative advertising and have laws designed to uphold the truth and accuracy of advertising, to help protect consumers from deception and markets from unfair competition. The overarching principle of advertising law is that advertisers be truthful and non-misleading when making claims about their products and services. Failing to comply with the law could subject advertisers to costly lawsuits, civil penalties, and damage their brand's credibility when judges' decisions become public. Except in certain special cases, advertisers must be able to "substantiate" claims about their products and services with the evidence that the claims are true. The laws that govern claim substantiation are drawn from a variety of different legal practice areas and disciplines, which are to be given a centralized home in this book.

In addition to laws under governmental jurisdiction, advertisers are subject to standards and guidelines from self-regulatory organizations (SROs) such as the National Advertising Division (NAD) in the United States, Advertising Standards Canada (ASC) in Canada, the Advertising Standards Authority (ASA) in the United Kingdom, and the Advertising Standards Council of India (ASCI), to name but four. Certain television

Ruth M. Corbin, (Editor): Practical Guide to Comparative Advertising
Governing Laws, Network Standards, and Industry Self-Regulation, David Mallen and Tom Jirgal, Principal authors.
DOI: https://doi.org/10.1016/B978-0-12-805471-0.00002-9. © 2019 Elsevier Inc. All rights reserved.

networks and Internet search engines also issue requirements for advertising claims they are willing to broadcast.

Laws and regulations are then the source of remedies for competitors to advertising companies; competitors who are damaged by false or unsubstantiated advertising may bring lawsuits or regulatory complaints against advertisers.

Claim substantiation refers to evidence demonstrating that a claim is truthful and not misleading. Such evidence will be required for any legal or regulatory dispute resolution that may follow the dissemination of a comparative advertisement. But what kind of evidence is required? How much evidence is enough?

The burden of substantiating advertising claims is often made challenging by the flexible nature of claim substantiation standards. In the United States, for example, most kinds of comparative advertising claims are required by law to be supported by a "reasonable basis" for the claims. What constitutes a "reasonable basis" will depend on the specifics of the claim and can vary from industry to industry. Broad and unqualified claims of superiority will generally require broad support. Tailored claims that focus on a single product attribute and/or single product variant can generally be substantiated with more narrow support. In respect to "reasonableness," there are no special rules for comparative advertising claims, other than rules for other types of advertising claims. Of course, the support for a comparative claim is more likely to come under scrutiny than for other types of claims, particularly when it calls out a competitor by name or by clear implication. Such claims should thus be subjected to a higher level of scrutiny before being used.

An examination of the various laws, rules, and standards for claim substantiation will provide a useful starting point for developing a sound approach to plan and substantiate comparative advertising claims.

LAWS AND GOVERNMENT ENFORCEMENT

Although principles of fair and truthful advertising are similar for all countries addressed in this book, different jurisdictions have their own specific wording and nuances of the law, administered by their own legal institutions. Such institutions may be the ones to initiate action, even where no company or citizen lodges a formal complaint.

In the United States, oversight of advertising claims comes at the federal level from the Federal Trade Commission (FTC), an independent government agency charged with promoting consumer protection and preventing unfair competition. Attorney General offices typically perform the same function at the state level.

Under Section 5 of the FTC Act,[a] the FTC is empowered to prevent "unfair or deceptive acts and practices," including unfair methods of competition. Most states have similar laws that are often referred to as "Little FTC Acts." Other countries have counterpart governmental agencies that regulate advertising practices and help ensure fair competition among competitors. For example, in Canada, the Competition Bureau oversees the manner in which businesses compete. In the United Kingdom, the Competition and Markets Authority plays a similar role, working to promote competition to the benefit of consumers. Access to information about other country institutions in Asia, Australia, Europe, and the Americas is found at http://www.competitionbureau.gc.ca/eic/site/cb-bc.nsf/eng/00290.html,[b] courtesy of the Government of Canada.

An advertisement is considered deceptive in the United States if there is a material "misrepresentation, omission or other practice that misleads the consumer acting reasonably in the circumstances, to the consumer's detriment."[c] The FTC's Policy Statement Regarding Advertising Substantiation (the FTC Substantiation Policy)[d] sets forth the kind of substantiation an advertiser should possess to avoid enforcement under the FTC Act. According to the FTC Substantiation Policy, advertisers should have a "reasonable basis" for their claims (either express or implied) *before* disseminating those claims in public. The substantiation must align closely with the wording of the claim. Specific standards for the level of substantiation are left flexible and may vary in accordance with factors that include the following:

- the type of product being advertised;
- the type of claim;
- the benefits of a truthful claim;
- the ease of developing substantiation;
- the consequences of a false claim; and

---

[a] 15 U.S.C. § 45.

[b] As on December 2017.

[c] FTC Policy Statement on Deception, *appended to Cliffdale Assocs.*, 103 F.T.C. 110, 175–176 (1984) (hereinafter *Deception Statement* or *Deception Policy*).

[d] *Appended to Thompson Medical Co.*, 104 F.T.C. 648, 839 (1984), aff'd, 791 F.2d 189 (D.C. Cir. 1986), cert. denied, 479 U.S. 1086 (1987).

- the amount of substantiation that experts in the field would think is reasonable.

To determine what claims an advertiser should be ready to substantiate, the FTC considers how a "reasonable consumer" would interpret an advertisement. Reasonable consumers are the ones who have average intelligence, skepticism, and curiosity. According to the FTC, an advertiser should be able to substantiate each reasonable interpretation that exists. Advertisers are not, however, liable for wholly unreasonable interpretations of their advertisements:

> An advertiser cannot be charged with liability with respect to every conceivable misconception, however outlandish, to which his representations might be subject among the foolish or feeble-minded. Some people, because of ignorance or incomprehension, may be misled by even a scrupulously honest claim... A representation does not become "false and deceptive" merely because it will be unreasonably misunderstood by an insignificant and unrepresentative segment of the class of persons to whom the representation is addressed.[e]

However, the "reasonable consumer" is not universal across all jurisdictions. The Supreme Court of Canada has ruled that the interpretation of commercial messages is to be assessed from the point of view of what a credulous and inexperienced consumer would comprehend.[f] Some states in the United States also apply a "credulous consumer" standard under their so-called Little FTC Acts.

Although there is no specific test for determining when an interpretation of an advertisement is "reasonable," consumer survey evidence can be instructive. When there is a dispute about whether an advertisement is communicating a particular message, the parties involved sometimes conduct a consumer survey of the advertisement to see what messages consumers are actually taking away from the advertisement. FTC case examples illustrate that if less than 10% of consumers interpret an advertisement to convey a specific meaning, that finding might be "insignificant and unrepresentative" of how a reasonable consumer would interpret the advertisement.[g] In contrast, a well-conducted survey demonstrating that 20% or more consumers are taking away a particular message from an advertisement is typically considered to be significant.

---

[e] *Deception Statement* at 178.
[f] *Richard v. Time Inc.*, 2012 SCC 8, [2012] 1 S.C.R. 265.
[g] *Firestone Tire & Rubber Co. v. FTC*, 481 F.2d 246, 249 (6th Cir. 1973).

In thinking about the "reasonable consumer," the FTC also considers the target audience for an advertisement. Advertisements that are targeted toward a particular group will be judged based on the likely interpretation of a typical member of that group. For example, if an advertisement for a calcium supplement is geared toward women aged 65 years and older, the FTC will consider the advertisement from the standpoint of women of that age group and may take into consideration such factors as whether that demographic is likely to have particular knowledge about osteoporosis that would frame its interpretation of a calcium supplement benefit claim. Similarly, advertising that is targeted toward children receives heightened scrutiny based on the limited ability of children to interpret and understand it.

In addition to the general prohibition against "unfair or deceptive acts or practices," the FTC periodically issues "guides" that provide examples or direction on how to avoid unfair or deceptive acts or practices when making certain kinds of claims. These guides address a broad range of varying advertising claims, including:

- Guides Against Deceptive Pricing (16 C.F.R. Part 233)
- Guides Against Bait Advertising (16 C.F.R. Part 238)
- Guides for the Advertising of Warranties and Guarantees (16 C.F.R. Part 239)
- Guide Concerning Use of the Word "Free" and Similar Representations (16 C.F.R. Part 251)
- Guides Concerning Use of Endorsements and Testimonials in Advertising (16 C.F.R. Part 255)
- Guides for the Use of Environmental Marketing Claims (16 C.F.R. Part 260)

The FTC guides themselves are not legal regulations but administrative interpretations that nevertheless provide valuable guidance to the kinds of claim substantiation the FTC expects advertisers to possess to make nondeceptive advertising claims. In addition, many states with Little FTC Acts have specifically incorporated the FTC guidance into their own statutes that do have the force of law. Additional laws and regulations may apply to advertising for particular kinds of claims (such as sale pricing or guarantees), particular classes of products or services (such as pharmaceuticals, food, alcohol, consumer leases, credit, or travel services), and for claims targeted to children.

In addition to the regulatory oversight imposed by the FTC Act and Little FTC Acts, consumers may also bring lawsuits, sometimes in the form of a class actions, for false advertising. Additionally, in the United States, the Lanham Act[h] gives companies the right to sue if they are injured by any "false or misleading description of fact" that a competitor utilizes in its advertising.[i] Of all the parties who might take issue with comparative advertising, an advertiser's direct competitor is likely to be the first to notice objectionable content that uses or implies the use of the competitor's brand indicia. Competitors will be motivated to bring lawsuits against companies engaging in false or misleading advertising when such advertising is likely to cause them harm. In such cases, aggrieved parties can seek not only an end to the false advertising by injunction or court order but also damages and in certain cases attorneys' fees. The Lanham Act is also a basis in the United States for trademark infringement actions and has counterparts in other countries. When engaging in comparative advertising that specifically identifies or describes competing companies or products, advertisers must also be mindful of trademark law and other intellectual property considerations.

In summary, in addition to the array of laws in almost every country that could be relevant to advertising, there is yet a wider array of such laws when comparative claims are added to the mix. That is one reason why a legal expert accessible to the advertising team is a must.

TELEVISION NETWORK STANDARDS

Television broadcast networks of significant size and reach set their own standards for claim substantiation that advertisers must adhere to if they want to have access to their audiences. US-based networks such as NBC, CBS, ABC, and Fox have departments of broadcast standards and practices that scrutinize every advertisement that is shown on their networks. Advertisers must submit their advertisements in advance to these departments—often at the storyboard stage—and they must generally submit a written summary of the advertiser's substantiation for each claim that is made in the commercial. Network personnel review the substantiation submitted by the advertiser and judge that substantiation against standards that the network publishes. If the substantiation comes up short, the network may reject an advertisement altogether or require that it should be modified.

[h] 15 U.S.C. § 1051, et seq.
[i] 15 U.S.C. §1125(a).

Different networks have their own processes and standards set forth in writing.[j] Some are more encompassing than others. In addition to prohibiting certain types of advertising, the standards set out rules of advertising, depending on the product at issue, the claim made, and other elements in the commercial such as product demonstrations. By way of example, ABC has separate rules for the following types of products and advertising claims:

- Alcoholic Beverages
- Children's Advertising
- Contraceptive and Fertility Advertising
- Dietary Supplements
- Direct Response Advertising
- Dramatizations, Reenactments
- Endorsements
- Financial Advertising
- "Free" Claims
- Guarantees and Warranties
- Lotteries
- Medical Product Advertising
- "New" Claims
- Personal Care Products
- Pet Food
- Price and Value Claims
  The list goes on.

The network standards also address comparative claims specifically. CBS and Fox make clear that comparative claims are permissible as long as they meet general standards for claims substantiation:

> [CBS Television Network] accepts comparative advertising (i.e., commercials in which the advertiser conveys the comparative superiority, in one or more respects, of its product or service over a specific product or service of a competitor). The standards that apply to such commercials are the same standards that CTN applies to all commercials. Comparative advertising is acceptable if the claims, comparative and otherwise, are truthful, fair and adequately substantiated. There is no restriction on commercials that draw meaningful comparisons, however strong, with specific competing products or services as long as these comparisons are fair and adequately substantiated.

---

[j] See https://nbcuadstandards.com/files/NBC_Advertising_Guidelines.pdf (NBC); http://abcallaccess.com/app/uploads/2016/01/2014-Advertising-Guidelines.pdf (ABC); http://www.fox.com/sites/default/files/FBCADVERTISERGUIDELINESFINAL2013-2014.pdf.

ABC and NBC have far more detailed standards for comparative claims, right through to the technical details of statistical testing and required promptness of updating the substantiation for a claim when a product is reformulated. The standards followed by both ABC and NBC favor comparisons that are specific, rather than general, with respect to claims of product superiority.

In summary, advertisers anticipating the use of television media should be aware of the particular rules of different networks with respect to claim substantiation. The rules need to be kept in mind right from the time that claim testing is planned.

SELF-REGULATORY ORGANIZATIONS

The advertising industry supports SROs in many countries to demonstrate its commitment to rules and standards that may go beyond even its legal obligations. Dispute resolution processes offered by SROs are typically more cost-effective and timely than litigation. Besides offering a forum for resolution of trade disputes between companies, SROs in different countries may monitor advertising, initiate their own investigations, or give copy advice to advertisers to help them stay out of trouble. The International Chamber of Commerce has a Consolidated Code of Advertising and Marketing Communication Practice that has been the foundation of advertising codes in most self-regulatory systems around the world.[k] Comparative advertising is explicitly addressed in many country codes, with agreed-upon standards and guidelines for making and supporting comparative claims.

In the United States, the NAD of the Council of Better Business Bureaus reviews advertising for its truth and accuracy in response to competitor challenges, as well as on its own initiative as part of its advertising monitoring program.[l] NAD's purposes are to advance consumer confidence and trust in advertising, promote fair competition, and minimize unnecessary government involvement and law enforcement activity. Over time, the NAD process has provided a valuable mechanism for competitors to resolve advertising disputes without resorting to costly litigation.

---

[k] http://www.codescentre.com/icc-code.aspx, last accessed December 16, 2017.
[l] NAD is part of the Advertising Self-Regulatory Council (ASRC), an independent self-regulatory body created by the advertising industry and the Council of Better Business Bureaus.

SROs like the NAD exist in many countries around the world. In the United Kingdom, for example, the ASA entertains consumer and competitor complaints, as part of its role to maintain consumer confidence in advertising, by ensuring that consumers are not misled, harmed, or offended by advertisements. Under Section 3 of the UK Code of Non-broadcast Advertising, Sales Promotion and Direct Marketing (CAP Code):

> *Particular care should be taken when making comparative claims against competitors and marketers should ensure that when doing so, they objectively compare one or more material, verifiable and representative feature of a product or service meeting the same need or intended for the same purpose, while ensuring they don't mislead consumers about either the advertised or the competing product.*[m]

India's self-regulatory organization, the ASCI, states its comparative advertising policy simply as the necessity to be "fair in competition," "not derogatory," and "no plagiarism."[n] Its website gives public access to regulatory rulings, including those adjudicating comparative advertising disputes between competitors. The latter constituted about 10% of complaints adjudicated by ASCI in 2016−17, and lack of substantiation for competitive claims has been a frequent determinant of outcomes.

The SRO in Canada, ASC, also plays an adjudicatory role and, further, has published Guidelines for the Use of Research and Survey Data to Support Comparative Advertising Claims, providing guidance to companies that advertise in Canada.[o] Australia also has an active system of advertising self-regulation. The Advertising Standards Bureau rules on complaints from both competitors and consumers.

The European Advertising Standards Alliance is an umbrella organization that "promotes responsible advertising in commercial communications by means of effective self-regulation, while being mindful of national differences in cultures, as well as legal and commercial practice." Its Code of Advertising and Marketing Communication Practice explicitly references comparative advertising, with the following governing statement:

> *Communications containing comparisons should be so designed that the comparison is not likely to mislead, and should comply with the principles of fair competition. Points of comparison should be based on facts which can be substantiated and should not be unfairly selected.*

---

[m] CAP Code, rules 3.33, 3.34, and 3.35.
[n] https://ascionline.org/index.php/principles-guidelines.html.
[o] https://www.adstandards.com/en/ASCLibrary/guidelinesCompAdvertising-en.pdf.

SROs play both an adjudicatory role (when they review competitive complaints) and an investigatory or enforcement role (when they review advertising through their self-monitoring programs). Competitive challenges involve two parties: a challenger contending that advertising claims made by a competitor are false or misleading, and an advertiser who defends its own advertising. NAD and other SROs typically play the role of a "neutral," listening to the arguments of the parties, reviewing their evidence, meeting separately with each party, and finding either that the advertising claims are substantiated or recommending modification or discontinuance of the advertising.

Final decisions of the ASA, ASCI, and the NAD are published. NAD's decisions appear in NAD/CARU Case Reports, available by subscription, and announced through the issuance of press releases.[p] The NAD online archive, a searchable database of published NAD decisions, documents NAD's rationale in reaching its determinations.[q] The decisions also provide guidance and precedents for marketers, attorneys, and consumers regarding the expected style and quantum of claim substantiation for different advertising contexts. Canada's ASC does not publish its decisions.

NAD's, ASCI's, and ASA's published decisions provide invaluable insight into the standards an advertiser must meet for its substantiation to survive legal scrutiny. NAD decisions direct that to produce meaningful results for claim support, product testing should be conducted under consumer-relevant conditions, using accepted methodology and protocols, and should relate directly to what the advertising claims. (Detailed considerations are addressed in later chapters of this book.) NAD looks carefully at the reliability of the claim testing, including whether the study was specifically designed for claim substantiation, whether the study tested the actual products at issue, whether there is any regulatory authority governing the testing standards for the particular category of products under consideration, and the statistical significance of the results.

NAD also looks to see if the claim is the one about which the regulatory authorities have already articulated a position (e.g., with regard to disclosures, green marketing, testimonials, and dietary supplements). Although NAD does not apply an industry code (in contrast, e.g., to the United Kingdom where the ASA applies the CAP Code), it looks to

---

[p] *Id.* at § 2.9 (C).
[q] http://www.asrcreviews.org/asrc-online-archive/.

such authorities for guidance and attempts to harmonize its criteria. Examples of regulatory bodies with which NAD has sought harmony include the Food and Drug Administration, the Federal Communications Commission, and the FTC.[r] In considering the degree of deference to be afforded such other regulatory authorities, NAD examines each case and evaluates the intent of the particular regulations and agency findings in light of the reasonable expectations of consumers.

If a reasonable basis for the advertiser's claims is established, the burden then shifts to the challenger to demonstrate that the evidence provided by the advertiser is fatally flawed or that the challenger possesses more persuasive evidence of a different result. Notably, NAD does not make findings of deception or law, or that an advertiser has engaged in false or unfair advertising practices. NAD's focus is on the advertising itself, and its authority is limited to recommending that an inadequately substantiated advertisement must be modified or discontinued. An advertiser's production of substantiation showing a reasonable basis for its claims will not necessarily end the NAD inquiry. An advertiser may yet find itself subject to additional recommendations for preventing consumer confusion, such as providing additional qualifying information or clear and conspicuous disclosures. Such recommendations may well go beyond what is required by law.

INTERNATIONAL REGULATION OF ADVERTISING

Although this chapter has included reference to rules, guidance, and regulatory environment of selected jurisdictions (the United States, Canada, United Kingdom, and India), advertisers who engage in comparative advertising need to be aware of rules and guidance in any country in which they advertise and sell products. Although nearly all countries have some form of "truth and accuracy" laws with respect to advertising, there are significant differences in the enforcement mechanisms and regulatory environment in different countries.

Other nations do not necessarily share FTC's stated policy view that comparative advertising, when truthful, can provide important

---

[r] However, an advertiser's compliance with the applicable regulatory framework does not necessarily insulate the advertiser from review by NAD because the mission of the self-regulatory system—to uphold the integrity of advertising—is often distinct from a particular concern addressed by a regulatory scheme.

information and help consumers make better purchase decisions.[s] In certain countries where comparative advertising has been more recently introduced, there appears to be less tolerance for the more aggressive style of comparative advertising or the outright naming of competitive brands. For example, although comparative advertising is permitted in China, advertisers may not make direct comparisons that specifically name competitive products and companies; in Brazil, comparisons can only be made with products manufactured in the same year; and in Japan, comparative advertising is permitted but rarely used. The different regulations in such countries appear to reflect differences in cultural adaptation: the level of comfort that each jurisdiction has gained with this relatively recent advertising style. Multi country advertisers must therefore look not only to the laws of different countries to determine what is permissible but also to the culture of their marketplaces to assess what will be effective.

In Europe, certain countries, such as Germany, had traditionally avoided comparative advertising. This too has changed. Comparative advertising is permitted under a European Union (EU) Directive, which defines comparative advertising as "advertising which explicitly or by implication identifies a competitor or goods or services by a competitor."[t] In the EU, advertisers, who engage in comparative advertising, must ensure that their advertisements:

1. are not misleading;
2. compare "like with like"—goods and services meeting same needs or intended for the same purpose;
3. objectively compare important features of the products or services concerned;
4. do not discredit other companies trademarks; and
5. do not create confusion among traders.

Notwithstanding that comparative advertising is permitted and practiced in Europe, EU rules are relatively restrictive of advertising that disparages or degrades the products or services of a competitor. Indeed, rules against disparagement are prevalent in almost every country where comparative advertising is permitted, with disparagement often interpreted more broadly

---

[s] Federal Trade Commission, 1979, Statement of Policy Regarding Comparative Advertising, available at https://www.ftc.gov/public-statements/1979/08/statement-policy-regarding-comparative-advertising.
[t] EU Directive 97/55/EEC.

than has been the case in the United States. In India, for example, advertisers can engage in comparisons and say that their product is better, or the best, but they are not permitted to say that their competitor's goods are bad.

Thus, although nearly all countries have laws intended to prevent deceptive advertising, there are many discrepancies, internationally, relating to comparative advertising in particular and to the rules for enforcement. Unlike other areas of laws governing international commerce (such as trademarks and copyright), there has been relatively little attempt to harmonize specific rules and advertising codes across countries. The responsibility to examine local laws and regulatory guidance in each nation where they advertise falls back on advertisers. Appreciating the need to substantiate claims with locally relevant evidence is a good place to start.

SUMMARY AND CHECKLIST

This chapter has presented the range of overseers of fair and truthful advertising. They include legal watchdogs, media broadcasters, and self-regulatory forums established by the advertising industry itself. At the foundation of all the regimes is the promotion of truthful, fair, and substantiated advertising claims. Comparative advertisements attract the scrutiny of competitors and carry a higher risk of legal or regulatory attention being called to dubious or controversial claims. By way of checklist reminders:

☐ Readers should compile the relevant (1) laws, (2) regulations, and (3) media rules for their own jurisdictions.

☐ From the early stages of formulating a comparative claim for future advertising, advertisers should obtain legal advice on the requirements to substantiate it and any precedents in law or regulation that should be taken into account.

☐ Even when rules are scrupulously followed, advertisers should anticipate competitive push-back in one form or another and have a plan—before the advertisement is launched—to respond accordingly.

☐ International advertisers need to revisit each of the summary steps listed earlier for each country in which they intend to extend their campaign. Ensuring locally relevant substantiation for a competitive claim will always be a good investment.

CHAPTER 3

What's the Name of the Claim[a]

Sound decision-making starts with knowing your options. There are so many ways to say that your company is great compared to the less great. Certain versions of that message may have more impact on customers than others. Some firms that we know have tested up to eight different versions of claims to learn which wordings will receive the highest percentage of customer endorsements.

But there is more to the issue than customer response. Claims in advertising must be backed up by facts. Different kinds of claims require different levels of backup information. Where should one start? Assuming you are committed to a comparative advertising approach, the first step to forming a claim is to ask: "What competitive benefit do I want to promote?" Then, consult the tools available for how to say it. This chapter is a primer on the categories of claims, identified by names you will hear used in the marketing and advertising business. The chapter also provides guidance on how to avoid trouble when choosing to word a comparative claim in a particular way. British and American regulatory agencies offer published cases as examples that are largely consistent with laws or regulations in several other countries.

[a] This chapter is an expanded version of an industry advisory article published by *Vue* magazine, October 2014.

Ruth M. Corbin, (Editor): Practical Guide to Comparative Advertising
What's the Name of the Claim, Dr. Ruth M. Corbin, Principal author.
DOI: https://doi.org/10.1016/B978-0-12-805471-0.00003-0. © 2019 Elsevier Inc. All rights reserved.

THE PINNACLE NUMBER ONE CLAIMS

You may have read number one claims such as "Trojan Brand Condoms are America's #1 condom," "Find out why Hill's Science Diet is vets' number one choice to feed their own pets," and "Australia's #1 Pure Coffee Brand." The phrase "number one" appearing in a claim conveys the fact of being at the top of the competitive heap in some respect. Permissible use of number one claims is governed by regulations and laws. Google has also developed its own policy regarding such advertising claims. According to a recent web posting, Google does not accept advertising text containing superlatives such as "best" or "number one" unless there is verification by a third party clearly displayed on the advertiser's website. Third-party verification, according to Google's policy, must come from a person or a group unrelated to the site.[b] Google's policy in this respect is in line with that of self-regulatory bodies for the advertisement industry in many countries: number one claims require proof.

[b] https://support.google.com/adwordspolicy/answer/6020955?hl = en&rd = 1,   as   on August 2016.

Gathering proof requires first that any ambiguity be dispelled. Number one at what? Or for what? Britain's Advertising Standards Authority (ASA) provides precedents on those questions. In the absence of any other clear meaning conveyed by the context of the claim, the ASA generally regards number one claims as meaning best-selling or, by implication, the competitor with the largest market share. In 2004, a complaint was launched in Britain against Scottishjobs.com for its claim of being "No. 1 for Plum Jobs." The ASA concluded that the claim erroneously implied that the advertiser offered more vacancies than its competitors and declared the advertisement unacceptable.[1]

Summary Lessons From the Trenches About Number One Claims
- Objective, third-party evidence is required.
- The context should make clear the answer to the question, "number one for what?"
- In the absence of clarifying "number one for what?" published precedents support an implied meaning of "number one market share."

GENERAL SUPERLATIVE CLAIMS

Instead of using the words "number one," some claims rely on more general words such as "best," "top," or "biggest." These are subsumed in the general category of superlative claims. You may have seen superlative claims like "The world's leading insurance and financial services organization," "The greatest show on earth," and "The most enjoyable car-buying experience you'll ever have."

The more specific the factors in which the advertiser claims the leading position, the clearer the direction for what needs to be proven. Complaints may arise if the performance criterion on which the advertiser claims to excel is left ambiguous. The ASA upheld a complaint against Time Computer Systems Ltd., for its claim of being "The UK's Biggest Retailer." Time Computers defended the claim by saying it had the most stores, but the ASA regarded the claim as falsely conveying that Time Computers sold the most PCs.[c]

---

[c] ASA nonbroadcast adjudication: Granville Technology Group Ltd., October 13, 1999.

Kevin Nash Group advertised itself as "The world's leading carp company." Fox International Group, a competitor, complained to the UK regulator that Fox had higher sales than Nash. Nash replied that it had not intended the claim to refer to sales, but rather that it had the broadest range of carp fishing products, including baits and reels. The relevance of even this evidence was challenged by Fox. The latter pointed out that Fox had been voted the best company overall in the previous 2 years by the readers of the magazine *Total Carp*, implying that even a generalized view of what it means to be "the world's leading company" in its field should favor Fox and not Nash. Nash was required to delete its claim from its advertising materials.[d]

What if a superlative assertion is just a matter of prideful boasting? Can claims be successfully defended as being based on the advertiser's opinion? A customer of Stoneacre Motor Group complained about the UK company's claim of offering "The most enjoyable car-buying experience you'll ever have!" The complaining customer insisted that he had experienced no such stellar enjoyment. However, the advertising regulator permitted the claim. It believed that readers would interpret the claim as merely stating the advertiser's opinion. In a similar vein, Papa John's Pizza fought successfully through two levels of US courts to maintain its advertising claim of "Better Ingredients. Better Pizza" as a statement of opinion.

Proceed with caution nonetheless: turning opinions into advertising claims without substantiation is akin to the proverbial risk of waving a red flag in front of a bull, the bull being the competitor one dares to challenge. Also, regulators may question the absence of substantiation even where the claim is recognized as being merely a statement of the advertiser's opinion. In 2005, for example, the ASA upheld a complaint about the claim "The ultimate broadband experience." While accepting the advertiser's contention that it was a statement of opinion, the ASA noted other evidence of severe customer dissatisfaction with the service. On that basis, the ASA considered the claim to be misleading.[e] Advertising Standards Canada has a related clause in its code (clause seven)[f] indicating that representations of opinion by any particular group must be based on

---

[d] ASA nonbroadcast adjudication: Kevin Nash Group Plc, October 27, 2004.
[e] ASA adjudication: Bulldog Communications Ltd., March 2, 2005.
[f] www.adstandards.com, as on August 2016.

the genuine experience of individuals in that group and must not deceive the public in any way.

Puffing for personal pride might also escape a finding of being misleading if no one would be expected to take the statement literally anyway. Such is the status frequently accorded to Barnum and Bailey's self-congratulatory description of their circus as "The Greatest Show on Earth." Canadian advertisers should take note of the difference in law between the United States and Canada on the subject of puffery. In Canada, the scope for arguing that an advertisement claim is just puffery is narrower than in other jurisdictions. Only clear hyperbole (i.e., a claim not related to specific product performance that is exaggerated to the point that it cannot be reasonably relied on) could escape the requirement for evidentiary support. One could wryly but realistically note that a claim of being "Best in the history of the entire universe" stands a better chance of being allowed in Canada than "Best in the country."

Well-intentioned advertisers have sometimes gone astray in assembling evidence they believe will support their superlative claim. Claims of "highest sales" must remain true for a reasonable period of time and not be subject to seasonal swings. Sales measurements themselves should be able to withstand the scrutiny of their validity. For example, a count of "For Sale" signs on lawns as evidence of a market-leading position in real estate was rejected because not all vendors opt to have such signs on their property.[g]

Summary Lessons From the Trenches About Superlative Claims

- As is the case for number one claims, the advertisement should make clear the answer to the question, "Best at what?"
- For factual claims (e.g., "Most chocolate chips per bite of any chocolate chip cookie"), factual evidence is required.
- Some superlative claims may be defended as the advertiser's opinion, but should not fly in the face of other evidence.

TARGETED SUPERIORITY CLAIMS, EXPLICIT OR IMPLIED

A superiority claim means "better than" the named or implied others. It has proven popular among many industries' leading rivals: Pepsi and Coke, Bell and Rogers, Molson and Labatt, to name just a few of the head-on competitors who have been embroiled in long-standing comparative advertising programs.

Companies sometimes make narrow superiority claims even while conceding they are not number one overall; they strategically use the top company in a category to point out their superior value on a single dimension. Avis became famous for its modest disinclination to take on number one rival Hertz. "We try harder," said Avis, from its proud number two spot. Certain President's Choice brands in Canada were compared favorably in their advertising to the relevant leading competitor, yet offering a lower price point.

---

[g] ASA adjudication: Blakes Estate Agents, May 1999

Targeted superiority claims may not always name competitors explicitly. The ad contains an "implied comparison." A tongue-in-cheek Burger King advertisement showed the back of a customer at its counter with a long plain gray coat, but the distinctive floppy shoes and curly red hair of Ronald McDonald gave away the customer's identity. "Four bucks is dumb. Now serving Espresso," announced McDonald's in a billboard ad, clearly targeting the high prices of one or more major coffee-shop brands. The Canadian Federal Court case *Church & Dwight Ltd. v. Sifto Canada Inc.*[h] featured a successful complaint for misleading advertising against Sifto Canada by the makers of Arm & Hammer baking soda (formerly Cow Brand). In touting superlative benefits, Sifto's advertising did not name any competitor. But makers of Arm & Hammer baking soda claimed that it was named by implication, given its dominant market share. The court agreed. The lesson learned therein was that one can still invite trouble even in the absence of direct reference to a competitor. If viewers would reasonably infer the identity of the targeted competitor, the advertisement is considered comparative, and evidence to support the claim is required.

Market leaders have proven notably sensitive about competitors' advertising referring to them in any way. Their top position is frequently integral to their reputation and business strategy. Legal action is tempting when competitors appear to step over the line of fair comparison. But legal action can take its toll in cost and undesired publicity. Companies are increasingly opting for alternatives to expensive and high-profile lawsuits. Such alternatives may include negotiations between respective lawyers, meetings between rival CEOs, or confidential arbitration. This trend does not let companies making advertising claims off the hook. Objective

---

[h] *Church & Dwight Ltd. v. Sifto Canada Inc.*, 1994, 7314 (ON SC).

evidence supporting or challenging the claim remains an essential tool for any well-informed pursuit of dispute resolution.

Summary Lessons From the Trenches About Superiority Claims

- Evidence is almost always required to support the claimed superiority among the pertinent population.
- Targeted competitors are usually on high alert for advertising that they believe devalues their brand. Comparative advertisers should assume their competitors will complain.
- Avoiding explicit identification of the targeted competitor will not necessarily keep an advertiser out of trouble.
- Enlightened companies have found some success with cost-efficient alternatives to litigation, by way of negotiation or confidential arbitration.

PARITY CLAIMS

A parity claim means "at least as good as." It does not sound like much—until you unleash the thesaurus. Try "unsurpassed." "Unsurpassed" asserts that others might be equal (parity), but no one is better. It is a particular type of parity claim, sometimes called "top parity," meaning "at least on par with the best."

Examples of top parity claims that the ASA has approved, with supporting evidence, are "You won't find the same deal for less," "ZOVIRAX antiviral cold sore treatment, nothing works faster," and "Unbeatable treatment for headlice." Top parity claims should be safe from regulatory censure if the advertiser's product is demonstrably as good as that of its main competitors and if there is no reasonable likelihood that it would be interpreted by consumers as a superiority claim. A case in point is an advertisement by Advil that caused headaches for Tylenol when it supplemented a parity claim with an engaging visual that created an impression beyond parity. The voice-over to the advertisement said only that Advil was "like Tylenol" in its effects on the stomach, but the court overseeing a misleading advertising dispute found that consumers took away a different message:

> The perception that Advil is not merely equal but superior to Tylenol in the respect of adverse gastric effects, is obviously not attributable to the 'like Tylenol' claim itself, but to other aspects of the commercials such as the "rolling pills" sequence... [T]he commercial shows first an aspirin tablet with the voice-over "I used to take aspirin." Then the aspirin tablet rolls away and its place is taken by a Tylenol tablet while the voice-over continues "... or Tylenol." The Tylenol tablet in turn rolls away to be replaced by an Advil tablet as the sound track announces, "But today it's Advil.... Advil. Advanced Medicine for Pain.[i]

Based in part on the survey evidence submitted, the court ruled that "the overall impression given, and obviously the one intended, is that Advil represents the current state of the art, a definite advance over Tylenol." The advertisement was enjoined.

Parity claims, modest though they may appear, are sometimes the right strategic choice. For example, they may be beneficial for a new product entrant who is willing to go up against the leading brand, perhaps at a lower price, or with greater ease of access. Grocery and drugstore retailers frequently introduce house brands whose packaging emulates that of well-known brands—perhaps hoping to convey equivalent value. The Canadian Broadcasting Corporation's *Age of Persuasion* commentator Terry O'Reilly has described how excitement and distinctiveness can be injected into parity products by shining a spotlight on a key benefit, even when the benefit is shared by more than one company in the category.[j] Recall how Crest took the historical lead in communicating to consumers: "You care about reducing cavities, we care about reducing cavities." Consumers whose own priorities match the advertiser's message come to identify with the brand. In short, being on par with others need not mean you are boring.

---

[i] McNeilab, *Inc. v. American Home Products Corp.*, 675 F. Supp. 819 (S.D.N.Y. 1987).

[j] The broadcast is reproduced at http://www.cbc.ca/ageofpersuasion/episode/season-5/2011/04/15/season-five-all-things-being-equal-the-fascinatingworld-of-parity-products-1/.

There is a special case of parity advertising for which one final caution is in order. It concerns the use of a competitor's trademark in parity claims. Whether such use is permitted in law is not clearly decided and not internationally consistent. The outcome of disputes in different country courts has been fact-dependent and subject to judicial interpretation of local trademark laws. The controversies have come into prominence in cases involving imitations of famous perfume brands, the so-called smell-alike claims. Imitator products have been marketed for many decades. But when advertisements call out the famous brand name that the imitator resembles, trouble may follow. Such was the case in *L'Oréal SA and Others v Bellure NV and Others*.[k] The defendants in that case marketed perfumes intended to imitate the famous L'Oréal brands, TRESOR, MIRACLE, ANAIS-ANAIS, and NOA. The imitation perfumes were marketed with sufficiently different names and packaging so that no confusion was entailed. However, a direct reference to the L'Oréal brands being imitated was made by the defendants in their marketing. Following an analysis by the European Court of Justice, the English Court of Appeal found that the defendants, through their advertised comparisons, had been "free-riding" on the reputation of L'Oréal brands, amounting to taking unfair advantage of their distinctive character or reputation. The court's decision reminds us of hidden risks even in the relatively tame category of parity claims.

Summary Lessons From the Trenches About Parity Claims

- Top parity claims require evidence that no other product in the category is better on the promoted dimension. It may suffice to test one's product only against the main competitors, i.e., the brands that have the largest market shares, and collectively account for most of the market. Which collection of competitors are to be considered the "main competitors" is a matter for judgment and risk assessment.
- Top parity claims may require larger sample sizes than a superiority claim targeted against a specific brand, because top parity claims require testing against more than one competitor.
- Fact-based parity claims require technical evidence of equivalence on the dimension in question and would benefit by disaster-check consumer
  (Continued)

---

[k] *L'Oréal SA and others v Bellure NV and others* [2010], EWCA CEV 535.

(Continued)
research to ensure that consumers would not infer a benefit beyond the stated fact.

- Advertisers who name the "comparable" competitor in a parity claim, even when the claim merely delivers facts, need to ensure they do not run afoul of trademark laws.

CUSTOMER PREFERENCE CLAIMS

Preference claims are about consumer attitudes. A claim may be made that people prefer the cleaning power or the taste of one product over another. Sometimes, the claim can be numerically specific: "Eight out of 10 mothers of newborns recommend Baby-Choice Diapers over the next leading brand."

Market evidence for preference claims typically comes in the form of surveys. To provide sufficient support, a survey must be reliable and geographically broad enough to accommodate the possibility of varying tastes and preferences across different regions. For in-person surveys, evidence from four major markets has constituted sufficient geographic breadth in both Canada and the United States to be accepted by regulators. Fewer markets for in-person surveys may suffice if the product's trade area is regional. If a preference claim is limited to certain competitors, then only the products named in the claim need to be tested against the advertised product. If the preference claim is more general (e.g., "The one most consumers prefer"), then the claim should be tested against brands that take up the lion's share of the market. However, regulators have not explicitly stated what would be sufficient to constitute the lion's share.

Summary Lessons From the Trenches About Preference Claims

- Preference claims must be tested against the products named or implied in the claim. If all products in the category are implied in the claim, then all significant competitive products should be tested. What constitutes significant competitive products is a matter for judgment.
- The test should be done on the pertinent population—the one whose preferences are being described.
- Preference claims lose their validity when competitors make major changes to their products.

PERSONAL TESTIMONIALS

Preference statements by single customers are sometimes delivered through an enthusiastic testimonial of a celebrity, a respected professional (like a physician), or a credible everyday person that the audience would want to emulate. Because single testimonials carry the power of story-telling, they carry emotional power and the potential to give the impression that "you too can experience what I have experienced." Recognizing this power to influence, regulations in many countries require that even authentic testimonials must reflect the genuine and reasonable current opinions of the consumer group that the endorser purports to represent.

For example, a 2015 decision of the Federal Court in Australia recounted proceedings taken by the Australian Competition and Consumer Commission (ACCC) against the franchisor of the Electrodry Carpet Cleaning business, alleging that it was involved in the posting of fake online testimonials. (Electrodry, through its more than 100 franchisees in Australia, provided cleaning services for carpet, drapery, upholstery, and mattresses.)

The ACCC alleged that Electrodry made false or misleading representations by posting fake testimonials on the Internet, as well as inducing or attempting to induce its franchisees to post their own fake testimonials. They also alleged that Electrodry's conduct resulted in fake testimonials appearing on review sites generally trusted by the public, including Yelp, Google, and True Local. The testimonials, ACCC claimed, appeared to be from genuine clients but were actually written or posted by people associated with, or contracted to, Electrodry. This was in direct breach of Australian Consumer Law, Sec'n 29(1)(e). The ACCC was successful, and Electrodry's franchisor was ordered to pay AU$215,000 in penalties. The judge summarized his analysis in the following plain-talk terms[1]:

> All the testimonials were posted on online review websites and were easily accessible to Australian consumers of the kinds of services provided by Electrodry Carpet Cleaning. The authors of the testimonials were not consumers of Electrodry Carpet Cleaning's services, had not been provided with such services and had not, as consumers, formed the favourable opinions that the testimonials represented. The testimonials were intended to convey, and no doubt did convey, that they stood as an independent and critical assessment of Electrodry Carpet Cleaning's services. They were, of course, nothing of the sort.

---

[1] *Australian Competition and Consumer Commission v A Whistle & Co (1979) Pty Limited* [2015] FCA 1447; quoted passage at para. 43.

Just as with other forms of false or misleading advertising, the fabricated testimonials had the potential to mislead a large number of consumers, divert customers from law-abiding competitors, and generate a positive perception of Electrodry Carpet Cleaning that was based on falsehoods.

Summary Lessons From the Trenches About Testimonials
- The testimonial must contain only true statements.
- The person delivering the testimonial must be a product user.
- The message delivered in the testimonial must be supported by an adequate and proper test, or, in some highly personal circumstances (like weight loss), may rely on a perceptible disclaimer as to what a "typical" result would be.
- Payments to endorsers must be disclosed.

SENSORY-BASED CLAIMS

A sensory-based claim refers to the human experience of sight, sound, touch, hearing, or taste (e.g., "creamier-tasting ice cream," "softer towels with our fabric softener," "a sharper-looking picture with our television sets"). The wording of the claim itself can fit with any of the categories reviewed earlier. Then why separate out sensory-based claims for special consideration? They require testing materials that go beyond verbal reactions to an advertisement, captured by way of routine surveys. They entail particular controls on how the product usage experience is tested. Indeed, a separate literature has sprung up to deal with their special needs, and ASTM publishes a dedicated set of standards for sensory-based claim substantiation. More details will be discussed on their extra needs in Chapter 4, Foundations of Test Design.

FACTUAL OR TECHNICAL CLAIMS

Comparative technical claims are usually facts about a product's components or performance that exceed what another company offers. The basis for a factual claim can be readily verifiable public data, such as "the airline with the most daily flights between Toronto and Montreal."

The basis for a technical claim may be the factual results of a laboratory test or scientific study. Canada's Competition Bureau cautions that technical performance reports must not mislead as to the circumstances in which a claim has been tested. "If the superiority of [a] product is limited to a certain range of conditions, then any superiority claim should be clearly qualified to reflect that range. For example, if a brand of gasoline were to be advertised as producing better mileage than several competitive brands, and the claim would be accurate under highway driving conditions but inaccurate under city conditions, this limitation should be clearly expressed."[m] It may appear that technical claims need only be backed by technical research, but advertisers should be aware of the hidden need for consumer vetting. By way of example, Rogers Communications advertised that its CHATR cell phone service resulted in "fewer dropped calls." Competitors argued that even if the claim were technically correct, the difference in the number of dropped calls between CHATR and competitors was as few as 1 in 500—too trivial to be meaningful to consumers. A lawsuit against Rogers for its advertised claim ensued. In defending against the suit, Rogers did collect evidence that, it argued, demonstrated a difference that consumers would find meaningful. The dispute eventually concluded with a half-million-dollar fine imposed on Rogers. Of all the issues in dispute, the court chose to base its sanction on Rogers' failure to have conducted an adequate test of the truth of the claim *in advance* of making it. The decision offers one lesson above all others: first prove, then claim.

SUMMARY AND CHECKLIST

Comparative claims can be categorized based on type. The claims typology helps marketers to plan what they want to say to the world about their product and what they need to do to earn the right to say it. Special cases of sensory-based and technical claims require methods of substantiation more complex than standard verbal surveys. While a specific claim may not fit perfectly into one type or another, the categorization puts a spotlight on the evidence required to support the main thrust of the claim.

Eight types of claims have been presented in total: number one claims, superlative claims, superiority claims, parity claims, preference claims, testimonials, sensory-based claims, and factual or technical claims. They are vehicles of communication and tools of persuasion. Each sets its own mandate for evidentiary research.

[m] http://www.competitionbureau.gc.ca/eic/site/cb-bc.nsf/eng/00513.html.

The steps to a successful choice can be summarized as follows:

- ☐ Decide on the competitive benefit you want to promote.
- ☐ Choose from among the options for expressing it. Choose one or more options for testing that you anticipate will be supported.
- ☐ Decide on whether you will identify a competitor by name or not.
- ☐ Seek valid and reliable evidence of the truth of the claim or claims being tested, evidence that could withstand the scrutiny of an expert hired by your competitor, if an objection is later raised. The evidence should be tailored to the wording of the claim—as close to the "exact" wording as reasonable.

REFERENCE

1. Vaas S. Jobs agency falsified its CV. *The Sunday Herald* August 8, 2004.

CHAPTER 4

Foundations of Test Design

OVERVIEW

Making a comparative claim almost always requires substantiation. When the claim pertains to sales or other factual data, the facts must support the claim unambiguously. When the claim involves what people perceive or infer, or how a product performs, a sound research test is required. As discussed in Chapter 3, What's the Name of the Claim, the type of claim you intend to make, right down to the exact wording, should be decided in advance of the research. Differently worded comparisons may require a differently constructed test. For example, claims of "longer-lasting flavor," "doctors recommend this vitamin over others," and "cleans as well as the leading brand," all demand different research designs.

Planning the research should involve cross-functional team members who have a stake in the successful production of the advertisement. Team members may usefully come from the advertising agency, the corporate marketing group, the legal department, and of course the company's market research, consumer insights, or sensory science groups. Each will have a professional perspective to enlighten the many issues involved in research planning. Each has a specific accountability for the success of the advertising.

The overarching guideline for test design is that the comparison being made between competitive products or services be objective and fair. A "mindset" worth adopting in evaluating your own testing plan is to imagine you were a competitor targeted by the advertisement: would the test appear to be objective and fair from both parties' points of view?

The test itself should be built on established principles of scientific research. Consumer psychologists and sensory scientists have developed tried-and-true experimental design tools, readily adapted to competitive claim testing. The test should have a clear objective of measuring user or buyer perceptions along the dimension relevant to the claim. For example, where one product is claimed to be "better-tasting," then the objective of the research will be to measure the taste experience. The outcome

Ruth M. Corbin, (Editor): Practical Guide to Comparative Advertising
Foundations of Test Design, Rebecca Bleibaum, Principal author.
DOI: https://doi.org/10.1016/B978-0-12-805471-0.00004-2. © 2019 Elsevier Inc. All rights reserved.

of the test must be unambiguous—there should be no extraneous possible explanation for why any one product prevailed over others. If two coffee brands are compared to find out which one consumers find "better-tasting," then, for example, the two coffee drinks to be compared should be consistently served at the same temperature. Serving temperature should not be allowed to confound the results of which one has the preferred taste. The same is true of all other features of coffee preparation and presentation.

In all respects, the research must be unbiased: all products or services being tested must have an equal chance to succeed or fail.

There are numerous choices to be made along the way in research design. *There is no perfect test design!* The legal team should be consulted for advice on research designs that have been accepted in the past by courts and regulators and on the reasons why research has sometimes been rejected. Such advice will help to avoid ill-fated design choices.

The remainder of this chapter addresses specific decisions and design standards for claims-testing, one by one.

THE CONTEXT FOR TESTING A CLAIM

An advertiser would normally start with a claim in isolation. Is it factually true, at least in the sense the advertiser intends? For some claims, supported by unambiguous facts, an advertiser may decide to go no further than assembling the factual support. "Best on-time arrival of any North American airline," is one such claim, where the supporting data speak for themselves. (A caveat was discussed in Chapter 3, What's the Name of the Claim, cautioning that a truthful claim of factual superiority still needs to be meaningful to consumers. If, for example, the airline boasting "best on-time arrival" relied on differences of 15 seconds on a few overseas flights in the previous year, it could well attract complaints of being misleading.)

For advertisements where research support needs to be gathered, the claim may also need to be tested as it appears within the advertisement itself. With surrounding visual or audio content, or implied ideas, what messages do consumers take away? People's perceptions are decidedly context dependent.[1] The messages that advertisers intend to send are not always the ones that consumers receive. The visual components of an advertisement may imply something different from the words or the words may contain unintended nuances. An example given in Chapter 3, What's the Name of the Claim, for an Advil advertisement described a case where an entertaining visual with pills rolling away gave consumers the impression of Advil's superiority, notwithstanding the verbal message of mere parity. Television advertising from India promoting Heinz's nutritional "Complan" drink mix provides another example of how visual story cues can spin innuendo beyond the meaning of the spoken words. Complan was being positioned against Glaxo's comparable "Horlick's" nutritional drink product, both products being marketed as contributors to the health and strength of growing children.

The commercial[a] featured a mother (the "Complan mom") and her son buying Complan in a store, and then explaining to another mother buying Horlick's (the "Horlick's mom," also accompanied by her son) why Complan was the better nutritious choice for growing children. As shown in the Slideshare file of Nisha Choudhary,[b] the son of the Horlick's mom is depicted as shorter and chubbier than the other boy, and wearing nerdy-looking glasses; the Horlick's mom herself is in drabber-color clothes than the Complan mom and has less glamorous hair and makeup. The Delhi High Court, while not opposed to advertising puffery in general, found the advertisement to be disparaging to Horlick's and beyond the realm of permissible puffing.

---

[a] Available for viewing in its original Tamil language at https://www.youtube.com/watch?v5MbCjODnZCVk&feature5related, last visited December 2017.

[b] At https://www.slideshare.net/nishatheperfectangel/comparative-advertising, last visited December 2017.

An advertisement may also mislead as a result of the omission of material facts. Such was the case found by the European Court of Justice[c] in opining on a recent French television advertisement by retailer Carrefour Hypermarchés SAS, comparing the prices of 500 leading brand products in its stores with those of competitor ITM Alimentaire. While the price comparisons were factually true, Carrefour's comparisons were done between stores of different formats. For its own prices, it chose its Hypermarché stores, whose size and location offered great economies of scale; for its competitor's prices, it chose their smaller neighborhood supermarkets. (Both companies had stores of both types.) The court observed that advertising that omits or hides relevant consumer information can distort the objectivity of a comparison and thereby mislead.

In another case of images exaggerating or distorting facts, a comparative advertisement for Dove Deep Moisture Body Wash[d] included the claim "some body washes can be harsh," accompanied by an image of an unnamed competitive body wash wrapped in barbed wire. Its main competitor, Dial, was identified by implication and complained to the NAD. In its defense, Dove pointed to the scientific literature supporting its claim that body wash can potentially damage skin proteins and lipids. However, the NAD found

[c] ECJ, February 8, 2017, Case C-562/15, *Carrefour Hypermarchés SAS v. ITM Alimentaire International SASU.*

[d] Excerpted from the *New York Times*, June 25, 2013, at http://www.nytimes.com/2013/06/25/business/media/in-criticizing-rival-products-a-dove-campaign-is-called-unfair.html, last visited December 2017.

that the message conveyed was false and disparaging, because the visual images implied, for some consumers, that the competitor's product would actually cause harm to the skin. Dove's support for the literal meaning of the words was not enough to justify the message that consumers took away.

In summary to all these examples, a fact-based claim in the context of its advertising setting may deliver a takeaway message beyond what the words say. Surveys or other tests of consumer impression are required to detect them, using the actual advertisement or (if the advertisement is still under development) the storyboards or other materials that come as close as possible to the final rendition.

CRITERIA FOR SOUND RESEARCH DESIGN

Any research design is assessable on the three criteria of reliability, validity, and relevance. Reliability refers to the reproducibility of the results on a different occasion with a different sample of participants. Reliable results can be generalized to a broader population. Validity refers to the ability of the test to capture the "true-life" measures of interest. Relevance refers to the pertinence of the results to the issues in dispute when an advertisement is challenged. Meeting these collective criteria calls for decisions on the following factors is to be reviewed in this chapter:

- Selecting the pertinent population
- Choosing sample size and geographic scope (including the special case of expert or consumer-based sensory panels for technical claims)

- Test administration: for sensory/usage claims, choosing intercept, central location testing (CLT), or home usage; for verbal claim testing, choosing a phone, Internet, or in-person testing
- Experimental design: repeated measures on the same consumer or comparison between different groups for different products
- Questionnaire design: wording on claim question(s) and decision in particular on including a "no preference" option
- For sensory/usage claims, choosing and sourcing the product
- For sensory/usage claims, preparing instructions on how to present the products to participants, or on how to deliver to them for home testing
- Recording the data
- Record retention within and outside the final report

SELECTING THE TARGET POPULATION

Advertising claims are aimed at target consumers of the product or sometimes at purchasers of the product. It is members of the target population who should participate in the claims test, rather than members of the population at large. For example, a test of retirement home advertising targeted at prospective residents would not reasonably include young adults as research participants.

Tests of advertising have generally used one of two approaches to finding demographically appropriate participants. If past research has produced demographic proportions of users, then those known proportions can be used to design the population sample for testing. The sample of participants would typically be balanced proportionally to product users based on age, gender, geography, and product usage or purchase interest. The choices of which demographics to use are a matter of judgment and available data. If target population demographics are not known in advance, a demographically proportionate sample of the *general population* can be vetted to determine if they are product users (or prospective purchasers) and only those who qualify would participate in the test. With the latter approach, the selection process will determine the demographics of users.

We have referred so far to "product users" as the appropriate test population, but, depending on the claim, it could be product purchasers instead. A children's building toy advertised as one that "kids prefer over LEGO" would ideally involve children in the testing stage; "better value for money than LEGO" is a claim that would more reasonably be tested

on parents and other adults who buy toys for children. Testing an advertisement about prices of a 24-case of beer would likely require an audience of beer purchasers, whereas testing an advertisement claim about the
preferred beer for taste would likely require an audience of beer drinkers
(whether they are the ones who actually go out to buy the beer).

Three interesting case examples are given next, wherein choosing or
not choosing the right test population made a decided difference in the
outcome of a competitive challenge.

An advertisement for Ragu pasta sauce by maker Mizkan Americas Inc.,
claimed "We have a Winner! Consumers prefer the taste of Ragu
Homestyle Traditional over Prego Traditional." Mizkan's research test used a
geographically dispersed sample of all tomato-based pasta sauce users and followed all of the recommended protocols established by the ASTM document, E1958 Guidelines for Sensory Claims Substantiation. Campbell Soup
Company, the maker of Prego pasta sauce, challenged the claim and provided
its own results showing parity between the two sauces when testing preferences of users of so-called traditional pasta sauce (a subcategory of tomato-
based sauces). Campbell presented expert opinion supporting its choice of
the survey population. NAD disagreed. "While it would be improper to test
a particular flavor among consumers who specifically do not like that flavor,
this is not to say that a taste test cannot be conducted employing subjects
who like other flavors." Well-experienced with the strengths and limitations
of survey evidence, NAD observed the requirement for substantiation to be
"reasonable" not necessarily "perfect." Mizkan prevailed.[e]

MOM Brands Company held taste tests to support superiority
claims for its Malt-O-Meal cereals. Its claims included the following:
"National Taste Test WINNER Fruity Dyno-Bites Preferred Over Post
Fruity Pebbles," "MOM Oat Blenders with Honey & Almonds Preferred
Over Post Honey Bunches of Oats with Almonds!," and "National Taste
Test WINNER Cocoa Dyno-Bites preferred over Post Cocoa Pebbles."
The competitor, Post, argued that MOM Brands' consumer population
sampled for the taste test did not accurately reflect the consumers of the
product. The test subjects included only adults aged 30−64 years, even
though the majority of those who ate the product were younger than 30
years (including children in particular). MOM Brands unsuccessfully

argued that children are not the purchasers of the cereal products, and that the claim was targeted to actual purchasers. NAD disagreed, noting that certain of the claims referred explicitly or by implication to "taste."[f] The case is of added interest, because NAD was otherwise complimentary of the advertiser's efforts to design a sound research test: the outcome turned to a significant extent on the choice of population.

A case between Weight Watchers and Stouffer's, producers of "Lean Cuisine," provides another cogent example[g] Weight Watchers sued Stouffer's over advertisements that claimed that certain of Stouffer's foods were equivalent to the so-called "exchange" value of comparable Weight Watchers menu options. In other words, Stouffer's seemed to imply, Weight Watchers diet followers could readily substitute Stouffer's products and continue on their determined road to healthy weight loss. Both sides submitted survey evidence in respect of claims of confusion and misleading advertising. The court wryly observed, "As might be expected, each side's expert on market research came to a conclusion that disfavored the other." Stouffer's survey was given entirely no weight, due to methodological flaws. About Weight Watchers' survey, the judge said, "I accord [it] slight weight, with strong misgivings about its improper universe." The universe of the Weight Watchers surveys was defined as women between the ages of 18 and 55 years who had purchased frozen food entrees in the past 6 months and who had tried to lose weight through diet and/or exercise in the past year. The court's analysis is instructive: "The [Weight Watchers] studies did not limit the universe to consumers who had purchased a diet frozen entree, or who had tried to lose weight through diet as opposed to exercise; therefore, some of the respondents may not have been in the market for diet food of any kind, and the study universe therefore was too broad. Sloppy execution of the survey broadened the universe further when interviewers mistakenly included participants who did not qualify even under [the intended criteria]. For example, on some of the qualifying surveys, not all of the questions used to qualify participants for the universe were answered; therefore, it is impossible to discern whether the respondent fits within the defined universe. Flaws in a study's universe quite seriously undermine the probative

[f] Industry press release at http://www.asrcreviews.org/nad-recommends-mom-brands-discontinue-taste-test-claims-challenged-by-post-advertiser-to-appeal/, last visited December 2017.

[g] *Weight Watchers Int'l, Inc. v. Stouffer Corp.*, 744 F. Supp. 1259, 1272-73 [S.D.N.Y. 1990]; this case description excerpted from.[2]

value of the study, because to ... be probative and meaningful ... surveys ... must rely upon responses by potential consumers of the products in question. ... Respondents who are not potential consumers may well be less likely to be aware of and to make relevant distinctions when reading ads than those who are potential consumers."

The pertinent population for survey evidence should be defined by the research team in consultation with its legal advisors. The case examples just given make clear the need for legal advisors to help anticipate what a future challenger could criticize. Brand and marketing teams, holders of information about their target consumers, will be able to provide input on specific recruitment criteria to capture members of their consumer markets. Such criteria might include age, gender, region, brand usage, or other factors known to dominate in the target market. A special ancillary questionnaire instrument, known as a "recruitment screener," is developed to capture members of the pertinent population. The screener both qualifies certain people to participate (based on predetermined characteristics) and eliminates prospective volunteers who do not meet the qualifying criteria. The screener is a key document for the verification of statistical reliability and should be retained as part of the research record.

What could possibly go wrong in drafting a screener to identify the right people for the survey? A critic will look for the following:

- Any extraneous motivational statements contained in the introduction that incline prospective participants to guess at the "right" answers, or even lie, to be able to qualify to take part.
- Screener questions that give away the purpose or subject matter of the survey, allowing prospective participants to formulate views before they hear the substantive questions.
- Omission of nonobvious, but important, screening criteria. For example, if respondents will be required to look at package labeling, they should be asked if they have with them any reading glasses or contact lenses which they normally require when shopping.

CHOOSING SAMPLE SIZE AND GEOGRAPHICAL LOCATIONS

The best choice of sample size and number of interviewing locations is a matter for professional judgment, taking into account the desired level of statistical reliability and the available budget. Several organizations publish recommendations, all with a view to achieving a level of statistical reliability that will provide assurance to future users of the survey results.

Sample size principles are highlighted here, with more detail about sample size in theory and practice addressed in Chapter 5, Statistical Support—How Much Is Enough?.

According to statistical theory, any well-chosen sample over 30 is considered "reliable" within a certain margin of error. Deciding on the sample size is thus a matter of business comfort or of credibility for use in a dispute resolution proceeding. Dispute resolution cases have accepted samples as small as 150 (having margin of error about 9%) for a claim affecting a well-defined region of a country, to samples as large as 500 or more (with a margin of error less than 4%) for a national claim. A user's level of comfort may be affected by his/her risk tolerance, the diversity of the population being sampled, or the variability of preferences and products being compared. The more variability in the marketplace, the bigger the sample should be, for greater comfort that all different types of people have had a chance to participate.

The key to good sampling is obtaining a representative cross-section of the overall target population. Representativeness is assisted by sampling *randomly*. Random sampling means that everyone in the population of interest gets the same chance of being selected.[h] A perfect random sampling of a very large population is nearly impossible but can be subjected to "best practices" in the industry to gain favor with courts and regulators.

Sampling across the entire geography of interest is desirable. For a national claim, the geography of interest is the entire country. Internet or telephone surveys can reach across an entire geography of interest. However, for tests that require participants to take active part in a personal experience, like a taste test or product trial, it is not financially practical or even feasible to sample all cities and rural areas of a large geographic expanse. Hence, a decision on "number of geographic locations" to be sampled must be made. Three or four markets have been considered sufficiently representative for testing claims in Canada and the United States. Even though three or four markets cannot capture true random samples of the target population, particularly when sampled in shopping malls, courts and regulators have routinely accepted them as a "reasonable best effort." Whatever the number, the choice of locations themselves should take into account any significant factors affecting product performance, such as weather, humidity, or distinctive cultures. For

---

[h] Or more precisely, a known chance or probability of being selected so that you can later adjust for the desired balance of characteristics.

example, the French-speaking province of Quebec, a distinct society in Canada, is routinely included among the markets chosen for testing in that country (when the claim itself is intended to extend to Quebec).

Typical industry practice entails a minimum sample size of 300 across three or four markets; this combination of sample size and locations has proven acceptable to support many claims.[i] The need for more people or locations depends on nuances or implications of the claim for specific target populations.

Expert panels or small populations may call for exceptions to these minimum sample size guidelines. Expert panels for sensory-based claims are treated in a later section.

ADMINISTRATION: TELEPHONE, INTERNET, HOME USAGE, OR CLT?

Telephone or Internet testing is suitable for testing claims referring to what people prefer, what people or professionals recommend, or what products people currently use. Internet testing may also be suitable for evaluating the impressions created by print, radio, or television advertising.

However, for claims addressing a sensory experience, like taste, smell, or texture, or for claims about the experience of using a product, some form of in-person testing is necessary. Where should in-person tests take place—at a central location facilities (CLT) or in the participants' homes? A brief description of the two options is given next, along with the advantages that each offers.

CLT involves screening participants in advance and scheduling them to attend, or intercepting them and asking them to come to a particular location for the test or the survey to take place. The location may be a mall, a focus group facility, a commercial kitchen, or some other conducive testing facility. Whatever the facility, it should permit a high level of consistency of participant experience, with respect to factors like product preparation and temperature controls, lighting, product storage, ambient sound, air flow with no extraneous odors, and product presentation. With CLT, participants can be given an opportunity to try more than one

---

[i] ASTM International, Committee E18 conducted round-robin research ("How Many Markets is Enough"; unpublished) with over 1000 consumers and based on the analysis that found around $N = 300$ consumers across three or four markets was sufficient to support most claims. Advertising Standards Canada has published the guideline of a minimum of 300 consumers across four geographic regions.

product at a time, in a consistent or balanced order of presentation. CLT also makes it easy to remove any brand indicators from the test products.

CLT was the preferred option for testing a claim of French fry preference between Wendy's and McDonalds. A description of the controls that were made possible by CLT demonstrates the value of choosing this option. The research was conducted "double-blind" (neither the interviewers nor the respondent was told which was which), the serving conditions were identical, and the CLT facilities were equidistant from Wendy's and the McDonald's locations in each market. Wendy's claim that its newly reformulated natural-cut fries with sea salt beat out McDonald's in a nationwide taste test was supported: 56% of consumers taking the test chose Wendy's, compared to 39% who selected McDonald's.[j]

[j] As per statement released by Wendy's describing the results, the lower picture is excerpted from its online video advertisement, declaring its superior taste, available at https://www.youtube.com/watch?v = tSe2-4bOKWYi.

CLT has certain disadvantages, compared to in-home testing, which make it less suitable for some projects. It permits only a limited-time or limited-occasion trial of the products in question and may thereby limit the simulation of real-life experience. CLT also entails the inevitable "demand characteristics" of experimental situations, according to which consumers pay heightened attention to product details in order to appear cooperative and intelligent.

In-home usage tests, in contrast, provide the opportunity for users to evaluate products over an extended period, as they might do in real life. The product placement in the home could last from a few days to a few weeks or longer, depending on the research design and, perhaps, on the wording of the claim. Home-usage tests are appropriate when the claim refers to how consumers experience a product under normal usage conditions for an extended period of time. Their disadvantages include the much longer time involved—weeks versus days, compared to CLT. Also, variability is introduced by the differences from home to home, including potential influence from other family members. Variability means less consistency. Why, the reader may wonder, would a research team be willing to give up control over consistency of the product-testing experience by allowing consumers to take it home? The answer is that the idiosyncrasies of home usage may be at the very heart of the claim. "The toilet paper dispenser that will make your rolls last 50% longer" is a claim that needs a home-use test. A research team does not need to know any particular household's toilet paper usage: in whatever way and at whatever speed a household uses a roll of toilet paper, the research team just needs to know the effect of the new dispenser compared to the household's typical usage.

In summary, CLT allows scientific quality and test protocol controls over every stage of the process. Home-use testing allows for real-life experience, with certain quality controls still possible, but with extra requirements for oversampling, scheduling, and product delivery. The choice between the two methods depends on the specific claim being planned and on what consumer activities are required to test it appropriately and fairly.

EXPERIMENTAL DESIGN: REPEATED MEASURES OR SEPARATE GROUPS?

To support a comparative statement, a design entailing "paired comparisons" is frequently used. In a paired-comparison test, two products are presented either simultaneously or one at a time to the same participant for a judgment. The judgment requested might be about preference ("which one do you prefer?") or about sensation (e.g., "which is sweeter?"). The question should be worded objectively, so as not to imply that one answer is better or more expected than another.

To recap, the two common methods for presenting two products being compared are as follows:

- *Simultaneous presentation.* Both products are presented at the same time, and the participant makes the determination of superiority directly.
- *Sequential presentation.* One product is presented and evaluated or rated according to some criterion, then removed from view. The second product is then presented and evaluated or rated. The question of preference or superiority is asked after both evaluations have been completed or preference/superiority is inferred from the respondent's rating of each product.

As in all claims testing, the wording of the questionnaire and test execution must match the wording of the claim.

Simultaneous presentation is not typically the preferred approach. Differences between products are more likely to be detected when two or more products are presented simultaneously; however, sequential presentation is more representative of consumer experience and ensures that the product being evaluated is not confused with the others in the array. Whichever presentation method is chosen, scientific objectivity requires balancing the order in which the products are evaluated or presented, such that each product is evaluated in each order position an equal number of times. This quality control measure avoids the risk of "order bias." With order rotation applied, if "being first to be evaluated" gives an advantage to one product over another, then both products get an equal chance to be the first. If order bias does exist, then it is minimized by the balanced order of presentation.

In some cases, it may not be feasible for any given participant to evaluate both of the products in a comparison test. This could occur with products that require longer evaluation periods or repeated use to reveal the product benefits, such as dandruff shampoo, wrinkle creams, or floor polish. A so-called monadic testing is then in order, designed such that "like groups of consumers" evaluate each product following the same protocols. Then a statistical comparison between groups is made. The appropriate statistical tests generally require larger product differences or larger sample sizes to validate a statistically significant finding.

The same approach of "separate groups" is adaptable to the situation when there are multiple products to be compared. Participants in separate groups may be able to evaluate just one of the products in a monadic design or a subset of the products in a so-called "incomplete block experimental design." Interested readers should consult an experimental design text for the array of options when multiple products are involved. Each such option is attached to an established statistical test of significance. In all cases, the presentation of products, if sequential, must incorporate a balanced ordering. The chart below summarizes the options discussed in this section.

Basic Options for Product Comparison Tests

What Participants Experience	Monadic or Multiproduct Design	Statistical Analysis Rules
Simultaneous presentation of products	Multiproduct: Every participant tests more than one product	"Within-group" design, whereby everyone's evaluations of one product are compared to their own evaluations of the other product(s). Even though all products are in view, evaluations of each should be done in different orders for different participants to avoid order bias.
Sequential presentation of products	Multiproduct: Every participant tests more than one product	"Within-group" design, whereby everyone's evaluations of one product are compared to their own evaluations of the other product(s). The sequencing order, in which products are presented, should be systematically varied to avoid order bias.
One product per person	Monadic	"Between-group" design. Different groups test different products (one each). Groups should be similar in relevant characteristics, such as age and gender, so that results cannot be influenced by how the people themselves differ between groups.

CHOOSING, PROCURING, AND STORING PRODUCTS FOR TESTING

When making a claim about your product "versus other major brands," a decision must be made about what other brands to include in the test. The decision depends on the so-called market structure—which brands have dominating market shares and in what proportion. A typical industry practice is to choose the highest market-share brands that collectively represent between 50% and 85% of total market share—choosing a percentage toward the lower end of the range for a fractionated product/service category where market share is difficult to establish and toward the higher end of the range for a well-defined category with a few major players and readily available market share information.

Brands identified in a competitive claim should be available in the markets where the claim is being made. For example, if Bright-O Laundry Detergent is claimed in a local radio advertisement to be just as effective as the best-selling brand, the "best-selling brand" being referenced should be the best seller in the geographic area served by the radio station, unless explicitly described otherwise (e.g., as "America's best-selling brand"). Also, when competing products are sold in more than one format, such as different sizes, concentrations, and solid versus liquid, the products used in the claim-support test should be comparable in format to each other (again an exception may occur if dictated by the wording or qualifying language of the claim).

Procurement of products is a more important quality-control step than may be appreciated by people unaware of the ways in which the same manufacturer's same-brand product can vary in the distribution channels. Products should be purchased from stores right in the markets where they will be tested, similar to how a consumer would purchase the products, rather than being shipped from the manufacturing site. All products that will be subjected to comparative testing should be purchased close in time to each other and of similar shelf-age, and from the same stores when possible. Products should be well within their shelf life. Once purchased, all products, while awaiting the day of testing, should be stored as they would be by reasonably careful consumers.

A word is in order about testing a new product that has not yet reached store shelves. In such instances, the test product should be sourced from its commercial production facility, using the equipment and operating conditions anticipated for future product distribution. In rare circumstances, it may be even too early for a commercial production facility to have been established. Then a so-called "pilot plant" may need to be used, whereby companies are able to replicate future manufacturing capability on a smaller scale. It is then advisable to document why products are unlikely to be materially altered from the pilot facility to the future larger scale version or to support this with sensory analytical testing.[3]

Whether products are new or already established in their life cycle, the guideline for product procurement is that the tested products are typical of what consumers would likely purchase in their local marketplace.

PREPARING PRODUCTS FOR PRESENTATION OR DISTRIBUTION

With CLT, products must be prepared, presented, and evaluated under reasonably realistic conditions and in a manner consisted with typical consumer behavior. Each product to be compared should be prepared following the manufacturer's recommended package instructions. Each competitive product should be presented and evaluated in a consistent manner. Products should be tested unbranded when possible, that is, without the participants' knowing which brand is which. Written instructions about product preparation should be drafted and sent in advance to test supervisors. The instructions should cover not only how the product is to be handled at every stage but also how the task will be objectively described to participants and allow for double-blind presentations to the consumer.

Preparing products and instructions for home usage tests requires similar attention to consistency and detail. Products to be compared must be assigned in a systematic balanced way; instructions should be delivered both verbally and in writing; record-keeping requirements need to be strictly followed. Some researchers add to research users' confidence (and to the confidence of a court if a competitive dispute arises in the future) by calling or communicating with participating households to receive updates, answer questions, and confirm that the participant is following instructions.

DEVELOPING THE QUESTIONNAIRE—TO USE OR NOT TO USE A "NO PREFERENCE" OPTION

In claims testing, the ideal questionnaire addresses the claim as presented, no more and no less.

How should it be structured? For "impression" tests, where it is necessary to ensure that the consumer is not drawing a false or disparaging inference from the claim, triers of fact have shown a preference for a questionnaire to begin with open-ended questions that do not structure a participant's thinking too early. Then a two-or-three question "funnel" sequence might follow, with questions increasingly specific but never leading or loaded. Where two or more products are being compared, as in sensory tests, the key comparative question should be positioned at the first logical place available, after whatever questions or instructions set the stage for the participants' task.

Some companies want to get more out of their research investment by putting extra questions on the questionnaire, not directly pertinent to the claim test. There is a risk in doing so, should the matter go to a court or regulatory dispute resolution. In a dispute resolution setting, the answers to every question are "discoverable." If confidential marketing information is collected in those extra questions, it will end up being divulged to a competitor in the evidence record. In the worst case, if any resulting information contradicts answers to other questions in the survey or weakens a party's case in any other way, the competitor will be able to use that evidence to its own advantage.

If a company is willing to accept the risks and add extra questions, the extra questions should go after those directly addressing the claim test. Then at least it cannot be argued that the extra questions influenced answers to the claim-test questions.

Whether a "no preference" option should be included has been debated by social scientists for a long time. On the one hand, an option to essentially say "I don't know" or "I can't choose" is an easy way out for people reluctant to give an opinion in case it is "wrong." On the other hand, forcing a choice where none exists may inflate the ratings for one product or another.

The standards of some television networks, and the assessment of some courts, lean toward favoring the use of a "no preference" alternative. One strategy for analyzing the data is to omit the "no preference" answers and look only at the favorability of one product over another among those who actually have a preference. Beware of the risk entailed in this form of analysis: it may be that the "winning" product is favored by a small percentage of people only. Consider the following hypothetical example:

> *A taste test is held between SMARTIES and M&M's chocolate covered candies. 15% of people favour SMARTIES, 9% of people favour M&M's and 76% of people have no preference. The differences are statistically significant.*

Owners of SMARTIES might want to use the results to claim that "people prefer the taste of SMARTIES over M&M's." The M&M's competitor might object that the claim is misleading because the large majority of people actually have no preference. Other less extreme examples could be given to make the issue trickier than it may appear in this example. To guard against the risk of allowing a potentially misleading claim to slip through the statistical analysis, ASTM has recommended making appropriate disclaimers whenever "no preference" votes exceed 20%.

DETERMINING HOW TO COLLECT DATA

Survey data collection is now largely computerized, even where human contact is made between interviewer and participant. Telephone surveys have "computer-assisted telephone interviewing," permitting interviewers to enter data as they receive participants' responses. In-person surveys outside the home are also able to benefit from computerized data entry, by having interviewers read questions from a portable computer device and enter responses accordingly. In-person tests in central locations may even permit participants to enter data on individualized computers in boardroom-type settings.

Whenever interviewers play a role in surveys, it is advisable that they do not know the purpose of the survey or its sponsor. Otherwise, they may unwittingly convey to respondents what answers are better or more expected from their point of view. When in-person sensory tests are conducted, a double-blind method is advisable. According to double-blind testing, neither the interviewer nor the participant knows the brand identity of either product being tested. Nor should either of them know the client for whom the study is being conducted.

Self-completed surveys by the Internet have introduced much efficiency to research-gathering organizations. They easily permit the answering of verbal concept questions, reactions to print advertising, or (if the production quality is high) evaluations of television commercials. Courts worldwide have shown a cautious willingness to accept them as

support for expert opinions. Still, Internet surveys have known vulnerabilities, such as the inability to obtain a random sample, the possibility
that participants fake their responses, speed through questions without
paying attention, or "cheat" on their answers. Published guidelines are
available for addressing the vulnerabilities of Internet surveys to enhance
their validity and reliability.[4]

SPECIALIZED POPULATIONS FOR SENSORY-BASED TECHNICAL CLAIMS

Some claims are based on physical or technical characteristics relevant to
consumer value, such as "our butter melts more quickly" and "our toma-
toes have a less acidic taste." Claims like those are frequently tested
through small panels of specially trained people, who are able to evaluate
a product's factual sensory characteristics, such as flavor, fragrance, texture,
flavor, and after-effects. There are two standard approaches to technical
claims. One approach is to use expert panels.

Expert panel members may receive more than 100 hours in training on
the modality required, on intensity references, and on rating scales to
record their perceptions.[5] These experts are not necessarily likers or users
of the product category but rather people with a developed sensitivity to
certain sensory qualities. Differences observed by such experts may or may
not be observable by everyday consumers, or be considered consumer-
relevant by a trier of fact. By way of example, SC Johnson & Son, Inc., the
maker of Pledge furniture polish, challenged claims made by Procter &
Gamble (P&G) in advertisements for its SWIFFER DUST & SHINE
Furniture Spray. P&G claimed that its Swiffer product left "less greasy residue than the leading furniture polish" on "your wood surfaces." The NAD
reviewed the head-to-head testing submitted by P&G in support of its "less
greasy residue claim" and found that while there was a technical difference
in the amount of residue left behind, the testing did not address whether
the difference was consumer noticeable or whether they would consider it
as "greasy residue." Even the consumer-use test submitted by P&G did not
persuade the NAD that its claim could be supported, because it still did
not establish that consumers could detect the difference in the amount of
residue or that the residue was perceived negatively as "greasy." The NAD
recommended that P&G discontinue using the term "greasy residue."

The surrogacy of expert panels standing in for relevant consumers has
also been controversial. Church & Dwight Co. Inc., the maker of Arm &

Hammer cleaning products, took offense to television commercials for Clorox's FRESH STEP cat litter that claimed that cats prefer FRESH STEP over Arm & Hammer's SUPER SCOOP litter because FRESH STEP was better at eliminating odors. Clorox subsequently ran another advertisement that made a claim of odor elimination superiority against Church & Dwight's entire line of cat litter products. Church & Dwight sought and received a preliminary injunction, temporarily halting the Clorox advertisements until a full trial was held. In granting the injunction, the New York District Court wrote that Clorox's laboratory "jar tests," using experts, were "insufficiently reliable to meet the required legal standards," and that the commercial risked causing "irreparable harm" to Church & Dwight. "It is highly implausible," wrote the judge, "that 11 panelists would stick their noses in jars of excrement and report 44 independent times that they smelled nothing unpleasant."[k]

A second frequent approach to substantiating technical claims is the use of consumer descriptions, by panels of consumers who are likers and users of the product category. Under the guidance of a trained moderator, the panel of consumers acts as jurors, creating their own sensory language to describe their perceived sensations, out of which arise the measurement protocols. Data are collected by way of those measurement factors, replicated across the panel. The resulting data are analyzed statistically and yield a quantitative assessment of product similarities and differences. Perceived differences are thus "consumer-relevant" by definition.[3] The research report should explain statistical reliability, using this "repeated measures" experimental design. It entails a different statistical approach than the traditional single-measure "margin of error" of large survey populations.

In testing approaches for technical claims, as described here, sample size requirements may be more modest than those for random samples of the population, because the "sampling" is rooted in repeated measures and inter-rater consistency. The specially qualified panels may have fewer than 30 people and oftentimes have about 12 subjects (simulating juries). The research designs are reminiscent of small-sample experimental designs in the physical and cognitive sciences, which have given rise to tailored statistical treatments.

More detail on specialized forms of sensory analytical testing is beyond the scope of this book but is available in other expert authorities.[6] Combining physical/chemical testing with one or both of the methods of

---

[k] *Church & Dwight Co. v. Clorox Co.*, S.D.N.Y., No. 1:11-cv-01865-JSR, 1/3/12.

human sensation testing described here will create "convergent validity" should all methods produce the same outcome, and thereby be more bullet-proof.[1]

SUMMARY AND RECORD RETENTION CHECKLIST

This chapter has covered structured, scientifically sound research designs for testing claims. Where surveys are used, they need to be structured to yield "proof" to a statistically reliable level. The field of sensory claim testing has relied on smaller scale designs than those entailed in mass-market surveys. Sensory claim tests, used to establish technical "truth" (regarding, e.g., odors, textures, and sounds), have employed smaller groups of trained panel members with demonstrated sensory acuity in the category of interest. Advertisers relying on such panels need to assure themselves and others that the evidence will be sufficiently consumer-relevant to justify a claim of sensory benefit or experience for product users.

Design decisions require trade-offs and judgment calls. These should be documented at every stage. Stakeholders in the research results will benefit from the retention of records of all decisions, protocols, and materials used. In a dispute resolution setting, such records will be essential to establish the integrity of the evidence. Reflecting the decisions you have made along the research design and implementation process, records should be kept for each of the following, where applicable:

- ☐ Choice of experimental design that serves the test of the claim, exactly as it is worded
- ☐ Methodology, in-person (central location or at-home), telephone, or Internet
- ☐ Selection criteria for the test population
- ☐ Samples sizes for each subgroup relevant to the experimental design
- ☐ The process for recruitment, screening, and qualification
- ☐ Test locations and dates
- ☐ Questionnaires
- ☐ Data collection process
- ☐ For in-person testing, interview/test staff instructions, including instructions on product handling, preparation, and serving
- ☐ Product procurement and storage protocols

---

[1] An instructive article on convergent validity in expert evidence appears in Ref. 7.

☐ Product preparation protocols
☐ Participant instructions
☐ Coding instructions for open–ended responses
☐ Photographs of the products and an overview of what a consumer is served may also be valuable as part of the research record

Consistent with full disclosure, most categories of records are usefully included in Appendices to the final report. Where they are not included in the report, legal and regulatory authorities should be followed with respect to how long records need to be retained after the evidence has served its main purpose (Box 4.1).[m]

BOX 4.1 The Lawyer on the Test Design Team

Torys LLP Partner Andrew Bernstein, recognized as one of Canada's top IP lawyers, was interviewed on how he perceives the role of counsel in the technical process of evidentiary test design. He views his role, he said, as distinct from and complementary to the marketing and research executives. "I do two things when commissioning evidence. First, I rely on my experts to tell me what methodology will hold up to rigorous scrutiny. But then I use my common sense to make sure that the experts aren't over-thinking it. Sometimes they are too smart for their own good, and in the end you have to ask yourself what the non-statistical judge or jury is going to think about the whole thing."

---

[m] Refer, for example, to ASTM E-1958-16a, for more information on test design and record retention.

REFERENCES

1. Corbin RM. Context effects on validity of response: lessons from focus groups and complacent frogs. *Vue* 2006;10−14.
2. Corbin R. Knowing who should be surveyed in comparative ad testing. *Vue* 2012;22−4.
3. Stone H, Bleibaum RN, Thomas HA. *Sensory evaluation practices.* 4th ed. San Diego, CA: Elsevier/Academic Press; 2012.
4. Mishra H, Corbin R. Online surveys in intellectual property litigation: Doveryai No Proveryai. *Trademark Reporter* 2017;107.
5. Meilgaard M, Civille GV, Carr BT. *Sensory evaluation techniques.* 5th ed. Boca Raton, FL: CRC Press; 2015.
6. Lawless H, Heymann H. *Sensory evaluation of food: principles and practices.* 2nd ed. New York: Springer Press; 2010.
7. Corbin R, Isaacson F. Surveys on a tight rope. The convergent validity net. *Intell. Prop. J.* 2012;**24**:265.

CHAPTER 5

Statistical Support— How Much Is Enough?

THE EVIDENCE IS IN THE NUMBERS

Comparative claims require support, usually in the form of reliable, valid data arising from a survey or product test. Statistical analysis is applied to research data from a sample of people to determine if the results are sufficient to confidently draw a conclusion about the population as a whole. Before drawing such a conclusion, one needs to rule out the possibility that the results could have occurred by chance. What does it mean for results to "occur by chance"? There is natural variation among people; when sampling randomly, you might end up with a sample of people who are not representative of the population and may get a distorted view. Such a situation occurs, for example, when political polls, based on a random sample, make a wrong prediction. Statistical analysis sets a criterion for how strong a result needs to be to rule out the possibility that the results can be explained merely by chance. There can never be a guarantee that the result for the sample is true for the entire population, but one can set an objective criterion for having, for example, 95% or 99% confidence that the result will be true for the entire population. Statistics permits a common language among stakeholders in the research, marketing, legal, and media communities regarding the strength of support for a statement or claim.

There are many types of statistical tests. The right one depends on how the research is designed and measurements are taken. For example, if the performance of two vacuum cleaner brands is being compared, the appropriate statistical test depends on whether each research participant tests both brands, or whether there are two groups of participants, with members of each group testing only one brand. If the measurement to be taken is simply "which one is the better picker-upper," then a different statistical test is called for than if the measurement to be taken is "how many days did the vacuum cleaner's battery last before needing to be recharged?" The appropriate statistical test can be determined from the structure of the research; but whatever the statistical test, it will answer

Ruth M. Corbin, (Editor): Practical Guide to Comparative Advertising
Statistical Support—How Much Is Enough?, Dr. Christine van Dongen, Principal author.
DOI: https://doi.org/10.1016/B978-0-12-805471-0.00005-4. © 2019 Elsevier Inc. All rights reserved.

the same broad questions: "Are the research results significant? If yes, how confident are you that the results did not occur by chance?"

INTERPRETING A MEASUREMENT DEPENDS ON THE UNDERLYING SCALE

Statistical tests deal with numbers. But even these precise-looking symbols called numbers can mean different things in the context in which measurements take place. Before analyzing numbers, we need to know the "scale" for their interpretation. For example, the statement "Light bulb A lasts longer than Light bulb B" only tells you that the light bulbs are ranked in this order: Light bulb A ranks #1 in how long it lasts and Light bulb B ranks #2 (when only two are being compared). However, to say that Light bulb A lasts three times as long as Light bulb B lasts would give you more detailed information about how far apart they are. Rankings and Time Measurements are on different "scales." Four basic types of scales are nominal, ordinal, interval, and ratio and have been described as follows:

In nominal scales, numbers are just designators that categorize or identify. They have no natural ordering. Numbers on football jerseys are one such example. In the context of surveys, when survey data are entered into a computer, gender is often coded as "1" for men and "2" for women, or vice versa. Numbers on a nominal scale are just neutral identifiers for things or people that typically have word-names of their own.

Ordinal scales are the second type. They are used for ranking products or people or attitudes. For example, the order of finishing in a cycling race produces a ranking of who crossed the finish line first, who crossed it second, and so on, until the last competitor. The ranks can tell the order of finishing the race but do not say anything about the actual time differences. So "magnitude" is conveyed with ordinal scales but not in a precise way. In survey research, questions that ask people whether they "strongly agree, agree, disagree, or strongly disagree" are based on ordinal scales. "Number one" claims in advertising and other general claims of superiority are, by design, based on ordinal scales.

The third type is interval scales. In interval scales, the interval between each notch on the scale has the same magnitude or meaning. The example most frequently given is temperature scales. An increase of 5 degrees on a temperature scale means the same thing, whether you add 5 to a temperature of 23 or 43 degrees. But there is no fixed "zero" in interval scales. Fahrenheit and Celsius temperature scales are both interval scales, but their "zero" point represents a different state of coldness.

The fourth and final type among the basic scale types is ratio scales. Ratio scales include a true zero point. "Height" and "price" are both measured on a ratio scale. Basic counting arithmetic you learned in grade school treats numbers as being on a ratio scale. When polling results are reported in percentages ("Chancellor Angela Merkel wins 33% of the vote"), they use a ratio scale: It would be correct to say that her support was approximately double that of a candidate who received 16% of the vote. With ratio scales, the ratios or proportions you calculate have a real-world meaning. When agreement or preference scales are asked in survey questions, the ratios between scale values will seldom be meaningful. Suppose, for example, that consumers are asked their strength of agreement with the statement "Downy Fabric Softener leaves my clothes softer than Bounce." If I assign the number "1" to an answer of "Somewhat agree," a "2" to an answer of "Agree," and a "3" to an answer of "Strongly agree" (this is technically called a "Likert scale"), I have no statistical basic to conclude that a consumer who answers "strongly agree" is three times more certain of his agreement than the person who answers "somewhat agree" or that his clothes feel three times softer. Indeed, you can see this intuitively by realizing that the numbers 1, 2, and 3 were themselves arbitrarily assigned—I could just as well have put the agreement answers on a 7-point Likert scale where "4" means "neither disagree nor agree," and then assigned "5" to "somewhat agree," "6" to "agree," and "7" to "strongly agree." In this example, the numbers assigned to levels of disagreement or agreement would not be on a ratio scale, because the "0" point is arbitrary.

Identifying the scale for the numbers you collect in claims research is critical to using the correct statistical test. A statistician will be able to advise which test to use, but as a stakeholder in the outcome of the research, you need to be aware of the "quality" or scale of the measurement before giving direction to a statistician. The four scale types are summarized in the chart below, along with the names of some common statistical tests used for testing differences (and therefore likely to be relevant to comparative claims). The test names are included for the benefit of technical readers who wish to investigate how to proceed with them or for nontechnical readers to discuss with their statisticians. The list contains references to broadly applicable statistical tests and is not exhaustive.

A caution is in order: every statistical test is based on assumptions—assumptions about the overall population, assumptions about how participants produce the measurements they do, and assumptions about the objectivity of the overall research design. A statistical result is only as good as the assumptions that underlie it.

Scale Type	What It Conveys	Typical Statistical Treatment
Nominal	Categorization only	Counting up how many appear in each category. Tests are called nonparametric. "Chi-square" is a well-used statistical test
Ordinal	Ranking on some dimension that can be used in a comparison	Comparison of rankings between groups. "Wilcoxon Signed-Rank test" and "Mann–Whitney U-test" are two of the available methods for investigating whether a difference is "significant"
Interval	Differences of degree, where the differences have consistent meaning	T-test or Analysis of Variance
Ratio	Any meaningful arithmetic operation using the numbers, including ratios, like "three times faster," "twice as popular"	T-test or Analysis of Variance

THREE CRITERIA SAY IT ALL

Defending claims with statistical evidence can be summed up by three criteria of good science: "reliability" (reproducibility of the results), "validity" (the true capture of the objects of measurement), and "relevance" (results pertinent to defending the precise claim being made). These principles are consistent with guidance from the Supreme Court of Canada for the evaluation of survey evidence.[a]

The terms reliability, validity, and relevance have sometimes been defined or used differently in practice in different professional communities. They have even been interchanged by judges in court decisions. For purposes of this discussion, we use the definitions mentioned above. The inconsistency among different professional groups inclines us to

---

[a] Mattel, Inc. v. 3894207 Canada Inc., 2006 SCC 22.

recommend that formal submissions of evidence define how these terms are used for purposes of that submission to leave no communication gap between writers and readers.

While we confidently advocate scientific standards, the results must be straightforward to describe in layperson terms—they have to make sense. Particularly when there is conflicting evidence between two parties and when the opposing arguments get technical and obscure, judges are inclined to fall back on their common sense to choose between the two.

We add one final observation based on the experience of compiling this book with authors from different professional communities. Professional associations for different market sectors have developed their own best practice guidelines. Experts within distinct professional communities are in the best position to know the limits of practicality or feasibility for their product areas. For example, sensory testing faces different challenges than opinion measurement. Thus, while the pursuit of reliability, validity, and relevance underpins all sound scientific testing, experts across different disciplines may differ on best practices for achieving them.

A HYPOTHESIS SETS UP YOUR OPPORTUNITY FOR PROOF[b]

Science provides a long-standing process, called the scientific method, for obtaining "proof" to an agreed-upon level of confidence. The basic idea is to set up a so-called null hypothesis (such as "the products aren't any different") and then to investigate whether there is enough evidence to *reject* the hypothesis ("wow, my product is better after all"). How much evidence you need and what you can conclude from it are the subject of this chapter.

The hypothesis you set out to prove depends on the claim. In general, the actual claim (as per the list of claim types in Chapter 3: What's the Name of the Claim) tells you the hypothesis that needs to be proven. The "null" hypothesis that you start with is usually easy to construct by stating the opposite. For example, if you wish to prove that "our brand is recommended more often than the leading brand," then the null hypothesis would be "people don't recommend us more often than the leading brand." You then investigate whether there is evidence to reject that hypothesis and accept the alternative one as reasonably "proven"—the

---

[b] Proof in this context means "supported to an accepted statistical level, within a specified degree of confidence."

one that is the basis of your claim. By "proven," we mean "supported to an accepted statistical level, within a specified degree of confidence." Exactly what that "level" and "degree of confidence" should be is a matter discussed later in this chapter.

The claim you hope to prove tells you what kind of test design to set up and what questions to ask. A company's great ginger ale product might be advertised as follows:

More Australian ginger ale drinkers prefer the taste of Gin Fizzy to that of the leading national brand.

Or it might be advertised this way:

Gin Fizzy is Australia's most preferred ginger ale.

Though both claims sound similar, their substantiation requires different test approaches and different survey wording. The first refers to taste, calling for an in-person taste test between Gin Fizzy and the established leading brand, and also calling for a question that refers explicitly to taste. The second refers to "preference"; the survey question in that case should explicitly refer to which one is "preferred." The first requires comparison to only one brand—the leading brand. The second requires a comparison to many brands. The first requires interviewing ginger ale drinkers only; the second might reasonably include interviews with purchasers who buy it for their families but do not necessarily drink it themselves.

But the scientific method for proving either is the same: start with an assumption that the statement is not true and set up an appropriate investigation for proof of the opposite hypothesis—namely that the claim is true.

WHAT DOES STATISTICAL PROOF MEAN? NOT 100% CERTAINTY BUT A HIGH DEGREE OF CONFIDENCE

Since testing is done on only a sample of the population of interest—not on the whole population—the results of a single sample can never absolutely guarantee what is true of the population at large. But random sampling allows for a certain level of confidence that your results would be "repeatable" on other occasions with other groups of people or objects as the case may be. Social scientists usually aim for a 95% level of confidence. That means that an observed result that supports your claim might have happened by chance but only a 5% chance. That is, even if no difference existed in the population, you could still find a difference in a *sample* of the population 5% of the time. More likely than not (95% of the time or 19 times out of 20), the result for the sample would be true for the whole population. The 95% choice is not etched in stone—it is just the one most frequently used and the one that legal and media communities in the United States and Canada have been shown to be comfortable with.

Put another way, when you survey a sample of the population and get a particular result, you do not know that you have found the true answer for the whole population—but you do know that there is a 95% chance that your result is within the margin of error of a true answer for the whole population.

RELIABILITY IS AFFECTED BY SAMPLE SIZE AND METHOD OF CAPTURING A REPRESENTATIVE SAMPLE

Reliability—that is, likely repeatability of the result for other samples in the population—is affected by the size of the sample and the degree to which sampled members of the population are actually representative of the population at large.

People unacquainted with rules of statistical inference may intuitively favor larger samples. However, large samples by themselves do not necessarily yield more accuracy than small samples. The notorious 1936 Literary Digest presidential political poll in the United States is a continuing historical reminder of this point. Based on a poll of 2 million voters, the Literary Digest predicted an overwhelming win for Republican candidate Landon. Instead, Roosevelt won over Landon with 62% of the vote. As it turned out, the magazine subscribers and other people who were polled constituted a biased sample of the universe of voters. Magazine subscribers at the time, it turned out, were not a representative sample of the population. The sample was discovered to be woefully unreliable. Whenever it is impossible, practically speaking, to sample the whole population of people, occasions, or objects that a claim pertains to, then random representative sampling of the population, to the extent possible, is an essential element of reliability.

Random sampling from existing databases can be done by established processes. Random samples of telephone numbers are straightforward to obtain, by the so-called random digit dialing, whereby banks of assigned telephone numbers are used and telephone numbers within them are selected by systematically altering certain digits in the phone numbers. However, with the evolution of the telecommunications industry, people are able to block or ignore certain calls; hence, the numbers dialed can still be randomly selected but the people you are able to reach are less so.

Random sampling can be simulated in public locations, like malls or grocery stores, by choosing every nth[c] person who passes by. Again, the people who are found in malls or grocery stores at the time a survey is conducted and also willing to cooperate with surveys may not be a true representative sample of the population, so even random selection among them will not produce an ideal result.

Random samples by the Internet can be systematically done from large panels of people (typically in the hundreds of thousands) who have agreed to be on such panels for purposes of participating in Internet surveys. As for other survey pools, such people are unlikely to be representative in every way of the population at large. But Internet surveys have made progress on at least one front and that is survey cooperation levels. Panel members are standing by to participate. So when you design a random sampling process from panels, you are likely to approximate a random sample at least of that panel population.

Sampling for sensory testing involves inviting participants to a central location or placing products in their homes. "Random" sampling is less likely to be possible, but such tests can incorporate a purposeful plan to have a cross-section of relevant consumers, locations, and testing conditions. Such a plan is one of "stratified sampling" that can incorporate a degree of randomization from among a pool of qualified volunteers.

Every method has its pros and cons. None is perfect. "Best industry practices" supported by excellent quality controls are likely to find receptiveness by a trier of fact when survey evidence is submitted for dispute resolution. At the very least, by using a rigorously randomized method of selection from an available pool, a researcher can defend his/her study participants as being randomly sampled from *some* population. If there is no logical reason why that population would be biased toward one outcome versus another, then a trier of fact can use common sense to infer that the results have widespread applicability to at least some material part of the population.

After good sampling is assured, then sample size comes into play in assessing reliability. Selecting a larger sized random sample will not necessarily change the "best estimate" of the true population parameter obtained by the survey. It may only change the precision, that is, the size of the margin of error surrounding the estimate. In other words, there is nothing

---

[c] "n" can be any number that is practical.

qualitatively poorer about smaller sample sizes. Their precision can be quantitatively defined by their so-called margin of error. Margin of error refers to the predicted range around the sample result for where the "true" population figure would be if one could indeed get measures from the whole population.[d] When samples are larger, the margin of error is smaller.

Statistical theory calls for a minimum sample size of 30 for making inferences about large populations.[e] The number 30 is derived from the Central Limit Theorem in statistics. The Central Limit Theorem describes what happens when simple random samples are drawn from a large population many thousands of times (or theoretically, an infinite number of times). For samples of approximately 30 or higher, the distribution of sampling results takes on the shape of a normal or bell-shaped distribution. Once a distribution is known to be normal, there are standard formulas available for calculating margin of error. Margin of error decreases as the sample size gets larger. Larger sample sizes are more expensive to obtain. In other words, both modest-sized and large samples, if both have been obtained by sound randomization procedures, would generate a supportable "best estimate" of the population parameter. The estimates would differ only in precision. You buy the level of precision you can afford.

The theoretical minimum of 30 is intuitively unsatisfactory in the noisy world of marketing. Samples will never be perfectly random; populations will seldom match the perfect bell curve for the characteristic in question. Even in a world of perfect sampling, a sample size of 30 has a margin of error that would be unacceptable in practice: its margin of error can be as wide as a plus-or-minus 18% deviation from the result discovered in the survey. That means that if it is found that 52% of a sample of 30 people prefer your product, the true result for the population could be as low as 34% or as high as 70%. Quite different implications for preference! So if 30 is likely to be unsatisfactory, what is an acceptable sample size? There is no fixed answer. Empirical practices have evolved for addressing expectations of different forums—like courts of law, regulatory agencies, or television networks. A sample size of 400 has an intuitively appealing margin of error of just under 5%, at the 95% confidence level.

---

[d] The margin of error is tied to a degree of "confidence" explained earlier.

[e] See, for example,[1] "While it will probably come as a surprise to many, the sampling distributions for most statistics achieve reasonable stability with a sample size of 30 or so. Using a larger sample size does not necessarily change the estimate or make it more accurate. It only permits us to reduce the size of the confidence interval surrounding the estimates."

Technically, that means that the true proportion in the population of a given result will deviate at most 5% from the discovered parameter, in 19 out of 20 cases of surveying samples of that size. Empirically, it means that the survey result is likely to be within 5% of the true result for the population, a number with which most courts, regulators, and industry experts would probably be comfortable.

A caveat to any recommended sample size arises when the researcher needs to analyze separately certain subsamples, such as men versus women, different age groups, particular geographic regions, or one of many interproduct comparisons. The anticipated margin of error for sub groups may then need to be taken into account in deciding on overall sample size, if sufficiently reliable inferences are to be drawn from the responses of key subgroups. Three hundred to 400 per subgroup is likely to be impractical and likely unnecessary in cases where findings about the subgroups are subsidiary to the main issues. It is not unusual to find sub sample size choices in the range of 50−100 (within a larger overall sample) for analyzing differences between subgroups of particular interest.

The sample size analysis just presented has been based on an assumption of a large and diverse population. For smaller homogenous populations, such as oncologists or life insurance agents in Minnesota, the margin of error surrounding a sample statistic can end up being materially smaller, and smaller samples may suffice. Nontechnical readers will find easily available references for determining appropriate sample sizes for these smaller so-called finite populations.

Since acceptable margin of error is a matter of comfort, it is not surprising to find differences in opinion of how much is enough. Professional industry groups, like the ASTM, have developed rules of thumb about what is generally accepted practice. Case law suggests that for most applications, a basic sample size of 300−400 is sufficient. The regulatory agency *Advertising Standards Canada* has published guidelines recommending a minimum of 300 for large populations.[f] Television networks set their own rules for advertisers. The following table[g] contains guidelines from different authorities related to sample size and geographic distribution:

[f] http://www.adstandards.com/en/Resources/guidelinesCompAdvertising-en.pdf.
[g] This table has been compiled by Jon Purther of CorbinPartners, Inc., in 2015, with input contributed by Dr. Christine van Dongen.

Source	Sample Size	Confidence Level	Regions
Advertising Substantiation, 2015 (Marketing Research Association—MRA) http://www.marketingresearch.org/issues-policies/best-practice/advertising-substantiation-and-standards-conducting-research	Minimum 300 (minimum 100 for subgroup/control group)	95%	No reference
Reference Manual on Scientific Evidence, 2011 (Federal Judicial Center/National Research Council) http://www.fjc.gov/public/pdf.nsf/lookup/SciMan3D01.pdf/$file/SciMan3D01.pdf (Page 381)	No reference	95%	No reference
Advertising Guidelines, 2014 (NBC) http://nbcuadstandards.com/files/NBC_Advertising_Guidelines.pdf (Pages 25–26)	Minimum 300	95%	At least four regions; at least two markets per region
Guidelines for the Use of Research and Survey Data to Support Comparative Advertising Claims, 2012 (Advertising Standards Canada) http://www.adstandards.com/en/ASCLibrary/guidelinesCompAdvertising-en.pdf (Pages 5–6)	Minimum 300	95%	At least four regions
Advertising Standards and Guidelines, 2011 (ABC) http://www.nbcadsales.com/content/programming/pdfs/2011%20Final%20NBC%20Advertising%20Guidelines.pdf (Pages 10–11)	Minimum 250	95%	At least four regions

(Continued)

Source	Sample Size	Confidence Level	Regions
What IP Attorneys Should Know About Expectations and Costs for Survey Research (The TASA Group—Technical Advisory Service for Attorneys) http://www.tasanet.com/knowledgeCenterDetails.aspx?docTypeID = 1&docCatID = 13&docID = 43	Minimum 200–300	95%	At least four regions
Conducting Community Surveys—A Practical Guide for Law Enforcement Agencies, 1999 (US Department of Justice–Bureau of Justice Statistics) http://www.bjs.gov/content/pub/pdf/ccspglea.pdf (Page 14)	Minimum 200–250	No reference	No reference
Survey Methods and Practices, 2010 (Statistics Canada) http://www.statcan.gc.ca/pub/12-587-x/12-587-x2003001-eng.pdf (Pages 152–155)	Minimum 400	95%	No reference
National Advertising Division (MOM Brands Company) Case # 5782 (November 2014) (referencing ASTM) http://www.asrcreviews.org/2014/11/nad-recommends-mom-brands-discontinue-taste-test-claims-challenged-by-post-advertiser-to-appeal/	No reference	No reference	Minimum two markets in each of the four regions
Standard Guide for Sensory Claim Substantiation, 1998 (American Society for Testing and Materials) http://library.sut.ac.th:8080/astm/cd15082005/PDF/E1958.pdf	Minimum 300	95%	Minimum two markets in each of the four regions

RELIABILITY DEPENDS ON CONTROLLING "NOISE" AND OTHER NONSAMPLING ERROR

The pursuit of statistical reliability requires more than good sampling; reliability also depends on controlling the so-called noise of everyday life—the random variations in behavior and environments that get in the way of perfect replications of each experimental trial or survey interview. Since reliability refers to the likelihood of getting the same or very similar results if the test were held with a different sample on a different occasion, there must be a high degree of quality control in how a test is administered across all its participants.

Preventable inconsistencies in research produce what is more technically referred to as "nonsampling error." Where claim substantiation tests are presented by advertisers, competitors will certainly look for inadequate control of nonsampling error in how the tests were carried out. Indeed, judges themselves have had occasion to label survey evidence as "inadmissible" for its failure to account for nonsampling error.[h] It is difficult to eliminate nonsampling error entirely, but there are a variety of ways to reduce it.

Interviewer Training and Monitoring

Where human interviewers are involved, whether in person or by phone, interviewers should have no knowledge of useful versus nonuseful answers, so that they do not unwittingly send cues to participants. Detailed, documented interview instructions and training help to ensure that interviewers understand their role in the survey procedure and act according to consistent guidelines.[2] Monitoring by an expert or supervisor provides additional assurance of ongoing adherence to instructions and consistency of behavior by interviewers.

[h] In *Toys "R" US, Inc. v. Canarsie Kiddie Shop, Inc.*, 559 F. Supp. 1189 (S.D.N.Y. 1983), a survey submitted as evidence which measured consumer perceptions of children's toy and retail stores was rejected on the basis that the purpose of the study had been disclosed to interviewers, the expert witness failed to monitor the interviews, and the interviewers had not been properly instructed on how to administer the survey. In another trademark case involving Toys "R" US, this one in Canada, over a canned nut brand called "Nuts R US," a survey submitted as expert evidence was rejected, in part, because the expert failed to supervise any of the interviews nor provide interviewers with any sort of instruction in the execution of the survey: *Toys R US v. Manjel*, 2003 FCT 283.

Consistency

Besides their interactions with interviewers, each participant should experience the test situation in as consistent manner as possible. Presentation orders of products or statements to be evaluated should be systematically rotated to avoid order bias.[i] If food products are at issue, the products must be prepared and presented in the same way.

"Blind" and "Double Blind" Surveys

Neither interviewers nor respondents should be told the purpose of the survey. In technical terms, the survey should be "double blind" in as many respects as possible. In the context of comparative claim substantiation, products that are presented to respondents should have their brand identity disguised and labeled, if necessary, with a neutral code.

Pretesting of Questionnaires

Questionnaires should be pretested to ensure ease of administration by interviewers and ease of question comprehension by respondents. Pretests or pilot studies typically involve the recruitment of a small sample of between 30 and 100 representatives of the pertinent population. Any difficulties or sources of confusion can be addressed before the official survey takes place. Where no difficulties are encountered in the pilot test, the completed interviews may be used as part of the overall official survey.

Optimization of Response Rates

Where the response rate to a survey is very low, one cannot be sure that the captured sample is truly representative of the overall population. Some assurance can be taken by checking the demographics of the responding sample (e.g., age and gender for each region sampled) and adjusting it to be proportional to the population at large. Efforts should be taken to optimize the response rate. For example, for telephone surveys, callbacks should be made to numbers that initially rang busy or were not answered. Courteous persistence by interviewers to persuade people answering their telephones to participate enhances response rates. Offering to call back at another more convenient time is also useful for not losing a prospective respondent.

[i] This is a minor caveat to the previous reference to a "consistent" experience.

HOW MUCH SUPPORT IS ENOUGH?
TAKE IT CLAIM BY CLAIM

The guidelines below on matching evidence to a claim assume a national two-product claim test conducted with 400 participants across four markets, supported by a test that has adhered to scientific standards for reliability and validity.

A "Majority Prefers" Claim

When testing your product against a single competitor product, it is tempting to treat any level of preference above 50% as sufficient to make a claim that "the majority prefers...." However, don't forget about the sampling and nonsampling error. The margin of error for a random national sample of 400 is just under 5%. Conservatively assuming nonsampling error to be approximately equal to the sampling error, a "majority" claim should only comfortably be made when measured preference of your product is over 60% (i.e., 50% + 5% sampling error + 5% estimated nonsampling error).

As a perhaps quirky feature of statistical inference, a more extreme result than a mere 50 + majority is detectable with a smaller sample size (assuming a constant margin of error). To see this intuitively, suppose that you are planning to sample 400 people to determine their preference between ice cream and broccoli; among the first 350 people you sample, 300 prefer ice cream. You already have a large enough sample size to conclude, with 95% confidence, that the finding so far of "85% preference" (300/350) will hold up for the entire population well within a 5% margin of error. In other words, if you set the bar in advance of what percentage you wish to include in the claim and if the bar is more extreme than 50% ("majority prefers"), then you may be able to achieve that bar with a smaller sample size. This approach to statistical verification entails *first* estimating the true result for the population and then calculating the minimum sample size necessary to detect that result. The approach may prove more efficient when you already have enough information in advance to estimate the level of preference in the population and when the test you plan is just for the purpose of producing scientific evidence for the scrutiny of others. The approach has been frequently used in the sensory-testing research community and is documented in more detail in the *Annual Book of ASTM Standards*, Volume 15.08.

The "Our Product Is Preferred Over Competitor X" Claim

Even without a clear majority preference, a claim of preference can be made whenever there is a statistically significant difference between preference for your product and preference for the competitor's product. This could occur, for example, with the following range of preference percentages:

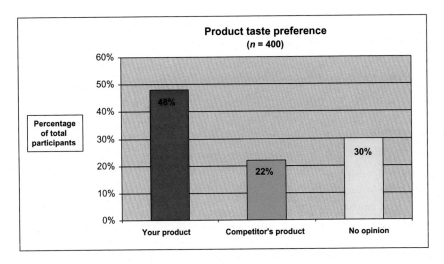

In other words, among those with a preference, your product is the clear statistical winner, but the "No preference" answers remove the option to make a "majority prefers" claim. Then you could still make a claim such as:

In a cross-country preference test between Gin Fizzy and the leading national brand, more ginger ale consumers preferred the taste of Gin Fizzy.

or...

Of ginger ale consumers stating a preference in a cross-country taste test, more preferred Gin Fizzy to the leading national brand.

A "just as much" claim a parity situation applies when the percentage preference for your product is statistically the same (within the margin of error) as that of your competitor's product. This could occur, for example, with the following range and percentage of preferences:

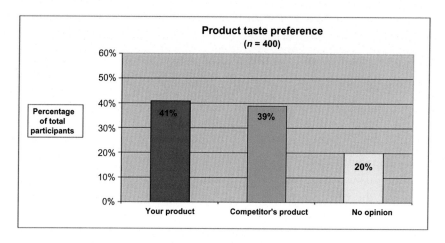

In this case, the claim can report on parity:

Canadian ginger ale consumers say they like the taste of Gin Fizzy just as often as the leading national brand.

or. . .

In a cross-country taste test, ginger ale consumers preferred Gin Fizzy as often as the leading national brand.

By way of a final word, the claim must always refer to, or unambiguously imply, the pertinent population from whom support was gathered. In all of the cases above, the audience in question was ginger ale consumers and the claims referred only to them.

STORIES FROM THE TRENCHES

The following stories reflect corporate experience; the names and some details are fictionalized:

Story 1: "Consumers prefer the taste of Homestead mayonnaise to Wonderful Whip.. . ."

In an effort to bolster the market share of Homestead mayonnaise, an ambitious marketer decides on a promotional message to reinforce preferences of existing purchasers and to help undecided consumers reach for Homestead on the grocery store shelf. Four hundred consumers drawn from four geographic regions in the United States were recruited and qualified to take part in a central location test. The sample comprised mayonnaise users whose last mayonnaise purchase was in approximate proportion to the market shares of Homestead and Wonderful Whip. Screening questions confirmed that they would use mayonnaise in sandwiches, salad mixes, and dips, so that their mayonnaise preferences would be relevant to a wide range of consumers. Products for the test were bought from local supermarkets in each region, all with approximately the same best-before dates. Brand labels were removed before the mayonnaise jars were delivered to the testing sites, and the jars were coded on the bottom with one of two three-digit code numbers that would allow the data analyst to know which product was which once the test was over. But the on-site interviewers did not know which was which. The research was thus double-blind. For each interview, the interviewer was required to enter only the three-digit code of the mayonnaise that the respondent ended up preferring. A no-preference option was also available.

Two test locations were used in each region. Exactly one tablespoon of each of the two mayonnaise products was delivered to consumers in half a turkey sandwich on white bread. At each location, the order in which the mayonnaise samples were presented to participating consumers was systematically rotated to avoid order bias. Interviewers were required to keep careful track of the order in which each mayonnaise had been tasted by each respondent, so that a data analyst would later be able to accurately attach the preference response to one of the two mayonnaise products.

Before respondents tasted the sandwiches, they were told, "The only difference between the two sandwich halves you will taste is the mayonnaise that was used. You will then be asked which mayonnaise you prefer." Respondents tasted the sandwiches in turn and were asked, "Which mayonnaise do you prefer, or do you have no opinion?"

Of the 400 consumers recruited for the study, only 390 completed it acceptably. Among the other 10, 4 decided they did not have time to wait for the test to start, 2 withdrew because they did not like turkey, 1 took a phone call between tasting the two sandwiches, 1 ate only the insides of the sandwiches without the bread, and 2 were overheard discussing their impressions with each other prior to giving their responses.

Of the 390 whose responses were deemed valid, 30 said they had no preference. Two hundred and fifty-five preferred the Homestead-dressed sandwich and 105 preferred the Wonderful Whip-dressed sandwich. Of the 360 consumers showing a preference, there is a statistically significant difference preferring Homestead. The claim is therefore supported. The evidence met the advertising standard of the television network, and the marketer's advertisement agency proceeded to produce and broadcast a televised claim.

Story 2: "Consumers like the taste of Dolly Madison's Chocolate Cake more than the leading Brands of Chocolate Cake Mixes..."

Grocery stores in America's west coast region sold three major brands of boxed chocolate cake: Dolly Madison, Jefferson's Choice, and Martha's Best. Together, these three brands controlled 85% of the market share in the region. Mixing and baking instructions were almost the same. The Dolly Madison R&D Team had recently improved the cake mix recipe. Internal taste evaluations led them to believe they now had a product that tasted superior to the two main competitors. Management authorized a research budget to test the superiority claim "Consumers like the taste of Dolly Madison's Chocolate cake more than the other leading brands..."

The Legal Department said the wording of the claim was fine, if it could be unambiguously supported, and asked for the report on the claim support test to be sent to them in advance of running the claim. The Marketing Department had already discovered, through market research, that "taste" was the priority of a majority of consumers and confirmed that their 2018 advertising campaign would indeed be positioned on "best-tasting," assuming they had the basis to do so. The Advertising Agency said that they could deliver an effective and entertaining advertisement based on that claim. So all of the major stakeholders had been consulted and supported the test of the claim proposed by R&D.

The claim test involved 300 users of boxed chocolate cake mix. They were selected from six cities, two each from three different states in the region, the states that showed the greatest sales volumes of chocolate cake mix. Users of the three different brands of cake mix were selected to mirror the market share of each of the three cake mixes. Cake mix products for each test site were obtained from the grocery stores local to the test locations and put into generic white boxes with code numbers only. On each box, the brand-specific mixing and baking instructions were added.

This was to be a home-use test. Each participating consumer received one different cake mix box for three consecutive weeks. They each baked a cake a week. There was a systematic rotation of how often each brand was given out in the first, second, and third weeks. Such rotation controlled for possible effects of research fatigue or order bias. After baking and tasting each cake, each consumer filled out a "hedonic taste rating scale" containing 9 rating points, with rating "1" corresponding to "Dislike extremely," rating 5 corresponding to "Neither like nor dislike," and rating 9 corresponding to "Like extremely." Consumer ratings were obtained by telephone interviewers at the end of each week.

An average rating was compiled for each cake, with the following results:

- Dolly Madison Average Overall Liking: 6.8
- Jefferson's Choice Average Overall Liking: 6.4
- Martha's Best Average Overall Liking: 6.5

An analysis of variance statistical test demonstrated that, although Dolly Madison rated highest, the relatively small difference between its rating and the other two was not statistically significant. Lack of statistical significance arises when there is so much variability in the data that differences of that small magnitude could happen by chance. The R&D Team was beside itself with disappointment. They asked if the statistician could not use the 90% confidence interval instead, if participants with the most extreme ratings could be dropped (since extreme cases are sometimes treated as "outliers"), or if only the results from two of the three states could be used. None of these proposals could be legitimately justified. Planning for the comparative advertisement campaign was discontinued. Potential disaster was avoided.

SUMMARY AND CHECKLIST

The use of statistics removes ambiguity. Once the technical experts confirm the legitimacy of how statistics are used, based on sufficient sample sizes and quality controls, then statistical statements can be turned into readily understood consumer claims.

Your readiness to incorporate the industry's major guidelines for support can be checked from the following list:

☐ Decide on the claim you want to make.
☐ Set up the wording for the hypotheses that will determine the test design.
☐ Get input at this stage from the advising lawyer, regarding any issues in principle with the wording of the claim. If issues arise, either modify the claim or ensure that the issues are covered off in the research design.

- [ ] Establish the "experimental design" that will determine how the variables of interest get measured (e.g., one per person or will each person assess all of the products?) and what statistical test will be used; establish protocols for the quality controls that will help ensure that the underlying assumptions of the statistical test are met.
- [ ] Determine sample size, including subsample sizes, taking into account the precedents or published minimums of the authority that will be deciding the acceptability of the advertising (such as a TV station, court, or regulatory authority).
- [ ] List in advance all the steps to be taken to control nonsampling error.
- [ ] Recruit a representative sample, ensuring that the consumers who participate are the ones that the claim is about.
- [ ] Determine the quantum necessary for the claim to be supported.
- [ ] If support is obtained, assess every possible reason why an opponent might disagree. Decide on whether the risks for going forward with the claim are tolerable.
- [ ] Document definitions and assumptions underlying the statistical tests for production in the report.

REFERENCES

1. Jacoby J, Handlin AH, Simonson A. *Survey evidence in deceptive advertising cases under the Lanham Act: an historical review of comments from the bench*, 84 T.R.; 1994. p. 541.
2. Corbin R, Gill K. *Survey evidence and the law worldwide: a reference text for lawyers, jurists and social scientists.* Markham: LexisNexis Canada Inc.; 2008.

CHAPTER 6

Know Your Limits: Claims Have Boundaries

The claim is supported, the media buy is made. The communication of a comparative claim may need one last accompaniment: a disclosure of the limits of what is being claimed, if there is any ambiguity or if a lack of disclosure might mislead. The three limits to be discussed in this chapter are:

- What you can say?
- What must be added to what you say, and how it must be added?
- How long can you say it?

WHAT YOU CAN SAY

What the advertisement can claim is what the test supports. You cannot imply something else through visual elements, tone of voice, or innuendo. There are no well-defined guidelines for avoiding innuendo—critics and competitors know it when they see it.

Specifically:

- A claim about perception or attitude must be limited to the perception or factor that was tested and not be generalized to overall preference. If people like the taste of a brand of ice cream better than a competitive brand, the claim can be "people prefer the taste of our ice cream" but not "people prefer our ice cream."
- If a claim is about performance, only the products (product store keeping units, or SKUs) tested in the research can normally be alluded to or even pictured in the advertisement.
- If the claim is about price, only the size or formats of the products/stores tested in the research can normally be alluded to or pictured in the advertisement.
- A caveat to the two guidelines above is to be made when an advertiser wishes to compare its product to all of the competitors in the marketplace (as in "we're number one"). When an advertiser wishes to make such a claim, then all the products that predominate in market shares

Ruth M. Corbin, (Editor): Practical Guide to Comparative Advertising
Know Your Limits: Claims Have Boundaries, Dr. Christine A. VanDongen Principal author.
DOI: https://doi.org/10.1016/B978-0-12-805471-0.00006-6. © 2019 Elsevier Inc. All rights reserved.

should normally have been included in the research—the ones with a smaller market share that "obviously" do not have a chance of being "number one" can sometimes arguably be omitted. How much of the overall market share the comparator products should represent is not a fixed percentage. The ASTM E-1958 claims substantiation document recommends a collective market share of 85%. In product categories crowded with many SKUs and fractionated market share, the testing of all products that make up 85% share of the market can be unfeasibly expensive. In such cases, an expert may be able to argue for which ones are clearly recognized as the only contenders for top spot.

In summary, the claim should not overreach to imply more, in a misleading way, than what the test has supported.

Certain regulated industries and professions, such as children's products, skin care products, pharmaceuticals, medicine, law, accounting, architecture, and engineering, to name but eight—are subject to their own advertising regulations, which should be consulted for additional rules about what can and cannot be claimed. Regulators for such industries bear an obligation for protection of health, safety, or vulnerable populations.

As for the content of the surrounding advertisement, there must be no unsupported or unfair implied comparison to a competitive brand. Seemingly neutral information cannot be added if its effect is to imply a meaningful comparison where none exists. For example, to promote canned corn as the only canned corn to have been boiled in pure spring water carries a meaningless benefit given that boiling tap water removes any relevant impurities.

Finally, there is the matter of good taste. Some network media add the expectation of "good taste" to what can be said on their stations. It is hard to predict what will be considered to be "in good taste" in any given year. Feminine care products have been singled out by some networks as products that should be depicted with care and sensitivity. Good taste also differs from country to country. Evolving social attitudes about women, relationships, and gender identity makes some advertisements more provocative or denigrating than others.

Consider the controversial advertisement for the Daihatsu Hijet van, bearing the caption "Picks up five times more women than a Lamborghini."[a] The fine print below the car elaborated that the car is "a

---

[a] https://www.reddit.com/r/AdPorn/comments/48yoa8/picks_up_five_times_more_women_than_a_lamborghini/.

babe magnet. . .it packs in six comfortable seats (four of them reclining). . . two sun roofs for when things get hot" with some additional double-entendre language. One can well imagine a different level of receptivity to this ad in different countries and communities. It certainly sparked controversy when it appeared, one website referring to it as "ad porn," another observing that it validated the objectification of women, while blogs can be found describing the humor as "brilliant."

WHAT MUST BE ADDED—DISCLAIMERS AND QUALIFIERS

Disclaimers are sometimes necessary to qualify a comparative claim, to disclose the demographics of the test sample, to identify the products involved in the test, or to reference the actual test results. Their importance may be underestimated by marketers who regard them as an afterthought. Disclaimers may provide essential contextual information about the limits of the advertised claim but cannot be expected to modify the general impression created by the associated headline. As effectively conveyed by two lawyers expert in advertising law in an address to the Canadian Bar Association:

> *Disclaimers are not magic bullets—they are simply part of the overall impression created. Insofar as they are less likely to be noticed or understood, or insofar as they are contradictory of the main message rather than simply providing additional detail as to what the main message is seeking to convey, disclaimers are unlikely to change an impression which is otherwise false or misleading in a material respect.*[1]

If a claim or statement would be deceptive in the absence of additional information, that information must be included in the disclaimer. Disclaimers must be clear and conspicuous. Even if the advertisement appears in social media or digital advertising where there is limited screen size, advertisers are required to include the qualifying information clearly and conspicuously.[2] The qualifying statement should meet the following criteria, which have been referred to as the "four Ps" of disclaimers.[3]

- *Prominence*: Qualifying information must be in large enough print for consumers to notice and read. It cannot be in proverbial "mice-type."
- *Presentation*: The wording must be readily understood by the average consumer.
- *Placement*: The disclaimer must be located where consumers are more likely than not to see it.

- *Proximity*: The disclosure must be near the claim it qualifies, to make the connection clear. An asterisk or footnote sign should direct the reader to the connected disclaimer. Examples:

"Raglos Curcumin....one soft gel, once a day -185X better"* [Footnote: *Compared to native curcumin extract.]

"Improves hydration of skin ...." [Footnote: *As reported by women consumers after seven days of usage]

In addition to the four placement rules above, there must also be attention to the content. Disclaimers and other fine print cannot contradict the impression given by more prominent aspects of the advertisement. Bell Canada paid the largest fine imposed by the Competition Bureau in Canadian history, 10 million dollars, when it announced a headline price for a new service and then referred to additional applicable fees in the footnoted disclaimer.[4] In a speech soon after the decision, the Commissioner of Competition said bluntly, in a public address "[i]ncluding a fine-print disclaimer is no license to advertise prices that are not available."[5]

Chapter 4, Foundations of Test Design, explained that offering test respondents a "no preference" option is recommended as a way to capture real-life possibilities rather than obliging respondents to make a choice where none exists. When communicating a preference claim, if a material proportion of consumers actually had no preference (the term "material proportion" usually considered to be in excess of 20%), a disclaimer should note that the stated preference is "among those who had a preference." If "no preference" votes are substantially more than 20%, it is easy to imagine a competitor challenging the integrity of the claim, whatever the disclaimer. Readers should be cautioned that the correct treatment of a "no preference" situation is an ongoing matter of debate in many quarters; the most recent published guidelines of the relevant regulatory authority should be checked before making a decision for the design of expert evidence.

Another sometimes essential topic for disclaimers is the source of survey respondents. This is especially the case when respondents were clearly not randomly chosen, such as when they have been derived from social media sites or a specialized panel. If an advertisement says, "Whiskey drinkers prefer Canadian Club...," and if the sample of whiskey drinkers was recruited only in bars, then the claim may not be supportable for those who purchase whiskey from liquor stores for consumption at home. Furthermore, if the whiskey drinkers were recruited only from the bars in urban business areas, this sample might be even less representative of the general population of whiskey drinkers. Indeed, because a disclaimer may not contradict a reasonably predicted impression of the main claim, it is arguable that the claim "Whiskey drinkers prefer Canadian Club. . ." should not even be made if its only support is bar drinkers in urban business areas.

Online panels and social media blogs are particular sources of nonrandom respondents but ones still rich in content. A company doing social media tracking was able to gather intriguing information that distinguished Dunkin' Donuts customers from Starbucks customers. Showing a branded cup of iced Dunkin' Donuts Coffee, the advertisement made the following claim: "Dunkin Donut drinkers like their coffee iced. They talk about iced coffee 3.7 times more frequently than Starbucks drinkers." A claim like that is possible with an appropriate disclaimer about the source of the data, being social media posts on specified sites within a specified date range, even though there had been no scientific survey. As long as the rules for gathering the data online were objective, relevant, and systematically applied, and disclaimers readily perceptive, a defensible comparative statement can be made.[b]

If a comparative claim is presented as a testimonial, the extent to which the speaker (like a doctor or opinion leader) is representing the views of his/her profession should normally be given.

HOW LONG YOU CAN SAY IT

A rule of thumb adopted by industry is that a comparative claim is likely to be usable for about a year ($\pm$ 6 months), assuming the market share stays stable. The US television networks have explicit published guidelines. Whatever the estimated shelf life of a claim in a stable market, its promotion would likely need to change if the comparator company changes its product; at that point, the comparator claim may no longer be true. Put another way, a claim cannot be defended as applying to the product that the competitor "used to have."

---

[b] https://www.slideshare.net/TrueLens/starbucks-dunkin2/18-DUNKINDONUTSDrrLIKE_ THEIRCOFFEE_icedTHEY_TALK_ABOUTiced.

SUMMARY AND CHECKLIST

Even well-supported claims have their limits: limits in what can be said, what must be said, and how long the claim is available to be made. If a competitor acts quickly to modify its product, the investment in the advertising program may have limited time value. This should be anticipated for fast action if necessary. Sometimes, of course, a short window of time after a product is launched—even if the advertisement is ordered off the airwaves—is enough for a company to make the impact it needs.

- ☐ Ensure that the wording of the comparative claim matches the test that supports it and is not altered in nuance or implication by other content of the advertisement.

- ☐ Challenge and test the impact of any humor, hyperbole, or emotional appeals incorporated in the advertisement—any of which may be interpreted differently than the advertiser intends.

- ☐ Add disclaimers as required, following the four Ps of Prominence, Presentation, Placement, and Proximity. Make sure that the disclaimer is not adding information critical to understanding the meaning of the claim or information that alters the predictable interpretation of the claim.

- ☐ Timing of the advertising program may take its normal course, unless the competitor changes its product formulation. If the claim is no longer valid, a competitor will undoubtedly let you know. However, it may be prudent to monitor the truth of the assumptions about the competitor's product that underlie the claim, at scheduled dates throughout the planned life of the claim, so as not to leave the business scrambling to respond to an allegation of a false and misleading campaign.

REFERENCES

1. Musgrove J, Edmondstone D. *The shifting general impression of disclaimers*, Address to the Canadian Bar Association 2012 Competition Law Spring Forum: Best Practices in a Time of Active Enforcement; May 2, 2012.
2. Federal Trade Commission. *How to make effective disclosures in digital advertising.* Available at: <http://www.ftc.gov/os/2013/03/130312dotcomdisclosures.pdf>; 2013.
3. Fair LA. *"FTC update: Enforcement priorities and key cases" Speaker at National Advertising Division annual conference, "What's new in advertising law, claim support and self-regulation?"* New York City; September 29–30, 2014.
4. Commissioner of Competition v. Bell Canada. Bell Mobility Inc. and Bell Expressvu Limited Partnership (Consent Agreement); June 28, 2011, CT-2011-005.
5. Aitken ML. Commissioner of Competition, Address (Keynote Speech delivered at the Canadian Bar Association 2011 Fall Conference, Hilton Lac-Learny, Quebec, October 6, 2011) online: Competition Bureau; 2011.

CHAPTER 7

An Ounce of Prevention: Troubleshoot Your Claim Before Launch

AN OUNCE OF PREVENTION AGAINST WHAT?

The lure of rewards from comparative advertising is seductive. You imagine a future where additional market share is yours, should the campaign succeed. Your company's product will stand out in consumers' minds from the myriad of competing products on store shelves. Consumers will finally be more informed about the wisdom of switching from the competitor's product to your own. Little wonder that companies have taken to making public comparisons. A 2013 survey of advertising executives found that more than 7 in 10 had created comparative advertising for clients in the last year, [1] about the same proportion since the last time such a survey had been done. Despite the various risks attached to all types of advertising, the author of the journal's report on the survey drily observed that "the greatest risk of comparative advertising may be the likelihood of a lawsuit." Large companies have not been dissuaded. Based on the continued barrage of comparative advertisements, they must be finding that the rewards outweigh the risks. The availability of sound research methodologies to test and substantiate claims surely calms the anxiety of advertising risk-takers.

Keeping their eye on the upside benefits of comparative advertising, marketers may be inclined to let the lawyers worry about the risks. That would be a mistake. Every professional in the company who is part of a comparative advertisement development process has an opportunity to mitigate risk for the benefit of the company and its shareholders. The US-based Paper Perfect Company that will be described in this chapter has hired Martin the Marketer, not only to grow market share in Paper Perfect's key product but also to manage market risks. We walk you

Ruth M. Corbin, (Editor): Practical Guide to Comparative Advertising
An Ounce of Prevention: Troubleshoot Your Claim Before Launch, Dr. Christine A. VanDongen, Principal author.
DOI: https://doi.org/10.1016/B978-0-12-805471-0.00007-8. © 2019 Elsevier Inc. All rights reserved.

through Martin's job and challenges in his first week. The steps he takes demonstrate the leadership role in risk management that marketers in similar roles have an opportunity to take.

On the first day on the job at Paper Perfect, Martin is advised that the Advertising and Sales Department will be meeting on Wednesday—2 days hence—to plan the company's first comparative claim for Paper Perfect paper towels. Martin's job is to develop the campaign strategy. This is his inaugural assignment for Paper Perfect Company—trial by fire! His MBA marketing courses taught him the theory and process for comparative advertising, but he has not yet learned much about the culture at Paper Perfect, the risk tolerance of its management, or the business style of its competitors. Martin sets out to educate himself quickly. He knows that a comparative advertising claim attracts attention and possibly the ire of high-profile competitors. He knows it may also fuel negative public reaction and competitive scrutiny—even to the extreme point of a legal action. He needs to assess the company's tolerance for risk—right up to the Board of Directors, whose very job is to manage the risks of the corporation. He schedules a meeting with the company's in-house General Counsel and legal advisor to the advertising department for 9 a.m. the next day.

ASSESSING STAKEHOLDER RISK TOLERANCE

Tuesday at 9 a.m., Martin meets with the company's General Counsel and in-house legal advisor to the advertising department. No company should proceed with comparative advertising without an advising lawyer. But at his former company, Martin was told by an in-house counsel, "This company doesn't do comparative advertising. End of story." It was indeed the end of the story at that company. A corporation cannot afford to proceed with a comparative claim without General Counsel in agreement. Without such agreement, even the Board should veto it.

Martin lays out the pros and cons of comparative advertising for Paper Perfect's three-ply paper towel product and explains why he believes that the risk is worth taking in this highly competitive category—that is, if the company actually holds the superior product. Paper Perfect's lawyer is receptive. As General Counsel to the Board of Directors, she assures him that she has the Board's authority to approve proceeding to the next step. She warns Martin that she personally has little time to look after his project and would have to hire outside counsel specifically to oversee it. Fees for outside counsel must be budgeted by the Marketing Department. How much? It depends on whether the competitor mounts a legal challenge. "Assume the competitor *will* complain," she advises. "That's always the right conservative assumption." "Would a legal budget of ten thousand dollars be enough?" Martin asks. The lawyer laughs.

ASSESSING FINANCIAL RISK

Martin retreats to his office on Tuesday afternoon to prepare a sober assessment of financial risk. The company already knows, from experience, what a standard advertising campaign would cost, based on its usual television buy. But developing an advertisement campaign with a comparative claim is bound to be more expensive. Martin contacts the colleagues who can give him informed figures for all of the following expenses:

- Internal and external lawyers' fees, over and above the costs routinely incurred for legal review. Martin anticipates a higher demand on his lawyers for advice and scrutiny when it comes to comparative advertisements.
- Costs for claim support research to be overseen by a research expert, with fieldwork contracted to an outside supplier.
- Internal and external costs of preparing a request for approval for television network review.
- Contingency budget for addressing direct complaints by a competitor.
- Contingency budget for regulatory or legal challenge.
- Addition of 20% on top of all figures, to set a high end for the budget range, in case of surprises along the way.

With a budget in hand, Martin prepares a well-supported analysis of the additional market share that can be expected by choosing a comparative advertising strategy and compares the value of additional market share points to the anticipated costs that would be over and above the costs of advertising that contains no comparative claims. This information gives him the business case to proceed.

INVESTIGATING THE COMPANY'S BEST BET

Martin's next stop is the R&D Department on Wednesday morning. He needs to learn what the technical experts know about product performance and consumer response to Paper Perfect paper towels. Martin is looking for his product's key meaningful benefit that a claim should address. He is also looking for any product issues that would immediately set back the chances of the claim's holding up. The R&D Department advises that it has just introduced a new three-ply paper formulation, with a quilted texture designed to hold moisture longer before seeping through the outer layers. Because the towels held onto moisture longer, the R&D staff explained, a big spill could be cleaned up with just one towel and no need to wring it out (at least in theory, based on greater absorption). Martin asks for any technical reports supporting the absorption power of Paper Perfect's new paper-towel formulation. He is pleased to find that a research report does exist, based on testing the new paper formulation with thick blue ink and monitoring the dissipation of the ink through the layers of the towels. The test was repeated for three other paper formulations that the R&D Department believed were in use by all other competitors. Competitive products varied in whether they used a one, two, or three-ply versions of one of those formulations.

Martin still has time before the Advertising and Sales team to check with the market research department and ask about the top product benefits reported by consumers. Market research staff handed him reports of consumer focus groups, in which consumers had talked about their confidence in Paper Perfect and their complete satisfaction in how well it cleaned up messes. The focus group reports contained entertaining stories about consumers' panicking when unexpected spills occurred and the speed with which they were able to remove all signs of the accident using Paper Perfect. Everyone in the focus groups seemed to agree on Paper Perfect's noteworthy speed of mopping up. Perhaps Martin could recruit one of the focus group participants to give an on-screen testimonial, he believes, but he could only do so if there were reliable quantitative data to support the general truth of the testimonial. In summary, none of the research on hand meets the evidentiary requirements of a claim about "speed" or any other product superiority, but the research at least helps to make the case for investing in a claims test.

Martin is optimistic that a superiority claim of some type could be sustained—something about Paper Perfect's superior absorbency. He postpones a visit to the Marketing Vice President until he is sure what the proposed wording will be. The advertising and sales team is meeting right after lunch. Martin joins.

The advertising manager starts off the meeting by suggesting three possible wording claims for a comparative advertisement: "The Fastest Mopper Upper," "Nothing absorbs better," and "Better than the brand you're using today—unless it's already Paper Perfect!" The sales team is brimming with enthusiasm for making a comparative claim. Salespeople regale Martin at the meeting with stories of grocery store managers being very satisfied with sales of the new Paper Perfect paper towel formulation. Still, they worry about competitive entries coming on to grocery store shelves at lower price points and higher margins for the retailers. They are convinced that consumers will remain loyal to the Paper Perfect brand because of its effectiveness in cleaning up liquid on countertops. The salespeople take turns in drawing Martin's attention to blogs from home-makers who share tips online, repeat purchase data, and a survey done 3 years ago by the marketing research department reporting high satisfaction ratings from their company's consumer panel, including satisfaction with the absorption power in particular. The meeting atmosphere is charged with optimism.

Martin is convinced by the enthusiasm and evidence of the sales team that a claim would be upheld of greater effectiveness than the competition. Still, none of the research or empirical evidence is right on point with regard to the speed of cleaning up. The technical absorption tests carried out by R&D cannot be relied upon because, first, they were done with blue ink, not with the normal variety of liquids that might be spilled in a kitchen (or other rooms in a home) and second, the difference in absorption may be technically true but not necessarily meaningful to consumers (if the difference in mop-up time is less than a second, say, would consumers even notice or care?). The focus groups were not done to scientific standard, and the quantitative consumer survey was done before the latest entry of a strongly competitive product. None of the research had performed a head-on-head comparative test with the paper towels of today's leading competitors. Whatever the claim to be made, it would have to be the subject of its own evidentiary test.

DECISION TIME: WHICH CLAIM TO TEST?

"OK, if R&D's ink test isn't enough," says the advertising manager, "just do it with water or juice, prove we're right about being best, and afterwards we'll decide on how to word the claim." Martin patiently explains to the advertising manager why they cannot proceed that way. Because he is new to the company and his colleagues do not yet have confidence in his expertise, his explanation is careful and respectful. "The three claims each call for a different type of test," he points out. "If we want to say our product is the 'Fastest Mopper Upper', then we have to measure speed of absorption of all the main brands. If we want to claim 'Nothing Absorbs Better', then we have to measure amount of absorption, again of all the main brands. If we want to make the claim 'Better than the Brand You're Using Today', we have to have each research participant determine which brand they'll compare with Paper Perfect; and we'll have to take a chance of defining what 'Better' means to consumers, and hope that viewers of the ad interpret it in the same way as we defined in our test." After a lengthy discussion of all the complexities involved in choosing a wording for the claim (who knew?), the team decides on "The Fastest Mopper Upper" as the claim they like best and think would be supported.

PILOT TESTING IS HIS MODEST COST FIRST STEP

To avoid unnecessary spending, Martin had first checked whether available data were sufficient to support the claim. Available technical tests, focus groups, and a survey done 3 years ago were all useful in instilling optimism. However, he knew that they would be insufficient for legal or regulatory evidence.

Claims-testing research is expensive, particularly where in-person trial with consumers is necessary (as in this case, where spill clean-ups are involved). Hence, Martin decides to start with a more modest cost pilot test in one city, that is, a smaller scale version of the full-blown survey. Pilot testing will allow Martin to learn about the likelihood of achieving the desired result before investing time and money in a broader scale undertaking.

A pilot test is initiated to test the claim of whether Paper Perfect is indeed the Fastest Mopper Upper (the wording of the claim is a bit quirky, but the advertising team believed that its humor and distinctiveness would stick in consumers' minds). The test will have to be designed to be able to conclude that Paper Perfect paper towel actually mopped up kitchen spills in fewer seconds than its competitors. Paper Perfect's three-ply formulation will have to be tested against other three-ply formulations (and the eventual advertisement will have to disclose that three-ply formulations were the ones being compared). The size of the paper towel will have to be constant across all brands tested. Consumers taking part in the test will have to notice differences in speed of mopping up, and a clear majority (after taking into account margin of error) would have to name Paper Perfect as the winner.

Next, a decision will have to be made about whether to do the test in a controlled environment or permit consumers to carry it out at home. There are pros and cons to both possibilities. Given the number of factors that have to be controlled (ordering of paper towels, timing measurement by seconds, capturing consumer impressions of speed immediately after mopping up, disguising brand identity, and consistency in the type of spills among all participants), a central location test is decided upon. Martin carefully documents why that trade-off choice was made, to be ready to respond in the future should an expert criticize the absence of "real-life home conditions." As he realizes the possibility that the research may indeed be criticized by an opponent's expert, Martin engages his own independent expert to endorse the questionnaire and research design and to take part in observing the research at every stage. The pilot test supported the claim. The roll-out to a national test supported the claim. Excitement was in the air.

TIME TO GO PUBLIC? ONE MORE STAGE OF DISASTER CHECKING

Statistically reliable support for the claim "Fastest Mopper Upper" was in place. An advertisement is designed showing a home fashion celebrity using Paper Perfect to mop up spills, and afterward running through every room in her home and tossing competitive products out the window. The celebrity then closes out the advertisement saying "Paper Perfect. The Fastest Mopper Upper. I won't have any other paper towel in my home." The marketing department is impatient to run the advertisement. It has sales goals to meet by the end of the year and every day counts. "Not so fast," Martin cautions, "let's pre-test the ad. Even though we have scientific support for the claim, we need to make sure there aren't any hidden problems in the ad itself". Martin's cautions seem obstructive to his marketing colleagues. The claim test was successful wasn't it? Martin presents them with many examples of advertising disasters that no one expected. A case in point was a New Zealand advertisement for Vittoria Coffee,[a] making the following claim:

[a] Advertising Standards Authority of New Zealand, Complaint 10/297 by Cerebos Greggs Limited, AWAP 10/006.

100% PURE COFFEE

It doesn't come in bags, it comes in bricks.

While other coffee companies will try to convince you that a soft pack is better, we all know that when ground coffee meets air it begins to stale.

That's why we remove the air in our vacuum-sealed coffee brick packs to guarantee the freshest ground Vittoria Coffee every time.

That advertisement, accompanied by a photograph of a coffee pack shaped like a brick, contained seemingly factual statements yet was found to mislead. As New Zealand's regulator observed, "the difference between coffee sold...in pouch or bag formats... is generally the variety of bean used...and manufacturing process. An ordinary consumer would understand the advertisement to imply that all competing coffee products sold in bags contain less than 100% pure coffee. The claim is therefore misleading and exploits the consumer's lack of knowledge of the manufacturing process for coffee." The advertisement had not been "disaster-checked" for possible interpretations not intended by the advertiser.

At this stage of the advertising launch process, focus group research may suffice. It is largely a disaster check against consumer interpretations that corporate employees may not anticipate. It is also a check for fairness in the eyes of others. Whereas fairness may not be well defined in law, it is a core principle of ethical advertising. Focus groups can be done among different audiences and in different regions. The focus groups will ask about the main messages that consumers take away; they will check on tone and on whether the advertisement is seen as mean-spirited or denigrating. Best to know before the launch.

PREPARE FOR BATTLE

The supporting research had been well done by Martin's in-house expert research team. However, trade-off decisions were made throughout of necessity and pragmatism, including the choice of central location testing over home use testing. Home use is highly variable, each household using the product in its own ways, thus decreasing the chances that a genuine product advantage will be detected. Martin sought out the most skeptical independent researcher he knew and could call on, to do a critical review of the claims research, and cite everything an adversary might point out. With an appreciation of the possible criticisms by a future adversary, Martin carefully documented and explained the reasons for each "trade-off decision" that had been made in the research design.

READY FOR HIS CLOSE-UP: MARTIN RELEASES THE ADVERTISEMENT

What can be expected on the day Martin authorizes the release of the advertisement through the media? When a comparative claim is advertised, the first to notice it is probably a competitor. Competitors will scrutinize not only the factual truthfulness of the claim but also the general impression left with a viewer. A "cease and desist" letter or phone call is the harbinger of trouble. If things can be worked out between the companies, all is well. But if not, there might follow a challenge by a competitor, consumer, enforcer, or self-regulatory organization (SRO), like the National Advertising Division (NAD), America's advertising self-regulator described in Chapter 2, Governing Laws, Network Standards, and Industry Self-Regulation. The NAD and other SRO's adjudicate disputes between advertiser and challenger and, after conducting an investigation and review, can make a recommendation—either for the advertiser to stop the advertising or for the advertiser to modify it to reflect only what can be supported. If the advertiser follows the NAD recommendation, the matter rests. If the advertiser does not stop advertising its product superiority, the advertiser may be referred to the Federal Trade Commission for enforcement action. Paper Perfect may also find itself in a costly legal battle if sued by a competitor or in a consumer class action lawsuit. The legal considerations facing Martin and his company, along with potentially substantial financial costs, are examined in more detail later in this book. While Martin may launch the advertisement campaign believing reasonably that his company is prepared for all battles, an exit strategy should be at the ready.

YOUR CHECKLIST FOR PROACTIVE TROUBLESHOOTING

- ☐ Know the standards of laws and regulations for fair and truthful advertising in your country or local jurisdiction. Look for case studies that may provide guidance for your company's own ambitions.
- ☐ Assess your company's tolerance for taking risks and ability to withstand challenges.
- ☐ Form a realistic and prudent view of the required investment to take your claim to public advertising, including resources for defending the claim if it is later challenged. Calculate the costs and rewards of

mounting a comparative advertising claim and ensure the financial analysis can support a business case for proceeding.

☐ Examine research-on-hand to generate comparative statements that have the best chance of being supported.

☐ Prepare objective defensible evidence to support the claim (or abandon plans if support is not forthcoming).

☐ Disaster-check the advertisement for unintended messages that consumers may infer.

☐ Negotiate with competitors who complain. Allow a complaint to proceed to litigation only if the company is ready for hostile scrutiny of sensitive information.

☐ Have an exit strategy as part of your overall contingency planning, well before the advertisement is launched.

☐ Use the resources available online via the government websites, NAD, and get the ASTM's, AMA's, and the Television Networks' claim guidelines as resources to educate yourself about claim substantiation research and processes. Go to conferences and workshops where claim substantiation is the topic.

BOX 7.1 An Interview With Sensory Scientist Lauren Rogers

Q: What advice can you give to a company that is about to embark on testing one or more comparative claims that it might want to feature in an advertising campaign?

A: Testing is time consuming and expensive, so we have to go into testing with a sound idea of the claim we want to make and that we think will stand up to scrutiny. I have two "Challenge strategies." I recommend to interdisciplinary corporate teams when they meet to decide on the claim they want to test. By challenge strategies, I mean a form of "contrary thinking" that helps to focus people's attention on assuring accuracy and reducing risk. The first is to ask team members to view a proposed claim as if it were being made by a competitor instead of us—that is, as if it were about a competitor's product. What kind of wording would we ourselves insist on, to ensure the competitor is not overstepping the line of fairness or accuracy? What kind of evidence would we insist they have? A second challenge strategy is to ask every team member to bring to the meeting a published case study with some relevance to our planned claim. The Advertising Standards Agency (ASA) in the UK regularly publishes its rulings at: https://www.asa.org.uk/codes-and-rulings/rulings.html. In the United States a similar listing is provided by the

(Continued)

BOX 7.1 (Continued)

Advertising Self-Regulatory Council: http://www.asrcreviews.org/. Reviewing cases where a company either won or lost in a dispute that relates in some way to our circumstances spurs people's thinking about how to avoid a future crisis.

REFERENCE

1. Beard F. Practitioner views of comparative advertising. *J Advert Res* 2013;**53** (3):313—23.

CHAPTER 8

Into the Fray: Playing Defense

INTRODUCTION

Every comparative advertiser should be ready to be challenged. It is true that many advertising claims are uncontroversial and the advertising campaigns that present them to the public go off without a hitch. Indeed, rather than thinking about the "other company," advertisers are more likely to focus on the business priority by promoting the brand and increasing sales or market share.

Advertising claims can, however, give rise to questions and challenges, by competitors, government enforcement agencies, or by consumers and the class action attorneys who represent them. An advertisement that is particularly aggressive or disparaging of another company or product practically goads adverse scrutiny. In the era of rapid electronic communication, it takes only a few minutes for one of your competitor's executives, agency staff, or employees who see the campaign to fire off a complaint to the corporate lawyers. In rapid succession may come a legal demand letter, a National Advertising Division (NAD) or network challenge, or even a lawsuit. It does not need a competitor to initiate action. Comparative claims, as with all claims, can also give rise in the United States to review from the Federal Trade Commission (FTC), state attorneys general, or other government agencies. To the extent they raise concerns for consumers, comparative advertisements can also result in consumer class actions.

The consequences of a challenge can be quite dramatic, and also expensive. Television networks may pull an advertisement if they determine that it contains an advertising claim that is not adequately substantiated. Additionally, a court can issue an injunction order for immediate removal of the advertisement if it determines that the advertisement is false and damaging to a competitor or is likely to be found so at a future trial. Investigations and lawsuits initiated by consumer protection agencies can give rise to significant financial liability either by way of a false advertising verdict or a settlement.

Ruth M. Corbin, (Editor): Practical Guide to Comparative Advertising
Into the Fray: Playing Defense, David Mallen and Thomas Jirgal, Principal authors.
DOI: https://doi.org/10.1016/B978-0-12-805471-0.00008-X.

In 2005, Gillette found itself on the receiving end of an advertising lawsuit when its competitor, Schick, alleged that the claims that Gillette made for its M3Power razor in a television commercial were false and inaccurate.[a] The advertisement claimed that the M3Power delivered "micropulses" which allowed the razor to raise hair up and away from the skin. Significantly, this measure of performance—raising hair up away from the skin—was not expressly stated in the commercial; rather, it was depicted visually using a computer-generated sequence.

According to the presiding judge, this depiction of hair lengthening was exaggerated and literally false. The court entered a preliminary injunction prohibiting the use of the television and print advertisements. Gillette was also ordered to change packaging for the product and remove in-store displays that featured the false claims. Additional lawsuits were filed by consumers in multiple districts resulting in years of litigation and settlement of as much as $7.5 million.

Even when an advertiser prevails in defending its advertising in court, litigation can be very expensive and disruptive to business. Every decision made in developing the claim, by *any* employee involved in the product development and claims-testing, could eventually be made a part of the dispute. In this chapter, we walk through the different types of legal challenges an advertising claim can be subject to and what is involved in each. We will also consider what lessons can be drawn about how to develop testing for claims substantiation purposes based on the scrutiny a test could face through a legal challenge.

EXPLANATION OF FORUMS

There are multiple forums where a company's advertising and its advertising claim support can be challenged. Understanding the different forums is useful for anticipating how evidence will be reviewed, the party with the burden of proof, the potential financial cost for the company, and the prospects for success. Where the following list refers to specific US-based institutions, readers outside the United States should investigate whether a corresponding forum exists in their countries.

- A competitor could bring a lawsuit challenging the advertising in court.

---

[a] *Schick Mfg, Inc. v. Gillette Co.*, 372 F Supp. 2d 273, May 31, 2005, reported in Advertising Age, "Court Rules against Gillette Razor Package Claim," June 23, 2005.

- A competitor could bring a challenge through a self-regulatory procedure, with the NAD of the Council of Better Business Bureaus, Inc.
- A competitor can advance a challenge of television advertising commercial to the network that carries it.
- A government agency at the federal, state, or local level may conduct an investigation of the advertising and the advertiser's evidentiary support.
- Consumers or consumer advocacy organizations can sue for false advertising—typically through a class action lawsuit on behalf of the consumer and others who purchased the advertiser's product based on the advertising claims alleged to be false.

The procedures, costs, and ultimate remedies in each forum are somewhat different, but most advertising disputes come down to two basic questions: (1) what message or messages are communicated by the challenged advertisements (whether expressly stated or implied) and (2) whether those messages are truthful and supported by a reasonable basis.

THE NAD

In the United States, the National Advertising Division of the Council of Better Business Bureaus, Inc. (NAD) is the most likely venue for challenging a competitor's advertising. NAD was formed in the early 1970s as a form of industry self-policing by national advertisers. A primary goal of the NAD is to increase the public's confidence in the advertising claims. Most of the advertising reviewed by NAD is initiated by competitors who utilize this system of voluntary industry self-regulation for the cost-effective resolution of advertising disputes with competitors. NAD sometimes engages in "monitoring" where it will open cases and review advertising on its own initiative. NAD's system of self-regulation is praised and supported by the FTC because it can also take some pressure off the commission and other regulatory agencies, whose resources are limited. (Government agencies charged with enforcing consumer protection laws tend to focus their enforcement efforts in areas of alleged fraud, health risks, or significant consumer harm.) As a result, comparative advertising is a particular focus of NAD disputes. Other countries have counterpart self-regulatory organizations, some of which play a similar role in the policing of advertising for truth and accuracy.

NAD is located in New York City and employs a team of attorneys with broad experience in advertising law and principles of claims substantiation. These attorneys hear and resolve disputes regarding advertising claims. As courts have become more crowded and the cost of formal litigation has soared, more and more companies have turned to NAD to challenge advertising claims by their competitors that they believe to be untrue in some material aspect. Presently, NAD typically addresses over 100 advertising challenges per year.

NAD's process generally involves written submissions by the two competitive parties and culminates with in-person or telephone meetings with NAD attorneys reviewing the advertising challenge.[b] A company that believes it has been harmed by a competitor's false or unsubstantiated advertising initiates the process by filing a written complaint, accompanied by the payment of a filing fee. The complaint needs to do no more than identify the advertising claims that the company is challenging, but typically a challenger will use the complaint to outline relevant information about the parties, the industry, the products at issue, and why the challenger believes that the challenged advertisements are making claims that are false, misleading, or unsubstantiated. Significantly, the burden of substantiating claims is on the company running the advertisement. As a result, the challenger has no obligation to include any evidence with the complaint regarding the truth or falsity of the advertiser's claims, but it may choose to do so if it would be advantageous to the case.

Once the complaint is filed, NAD reviews the complaint and opens a proceeding, unless there is a jurisdictional reason not to. NAD's jurisdiction extends only to advertising that is distributed nationally so it is unavailable as a forum for disputes about local or single-state advertising. NAD will generally decline to review advertising claims that were voluntarily discontinued prior to the initiation of a challenge. NAD's rules also preclude it from opening a new challenge in situations where an advertising claim is already the subject of court litigation, a court order, or an order from a federal agency.

NAD review is a voluntary process. No advertiser is legally obligated to participate in the NAD process or defend its advertising there. Most companies do so, however, because there are consequences if they do not. One consequence of refusing to participate is that NAD will issue a press release indicating that the advertiser failed to appear before NAD and defend its advertising claims. This could undermine the public's confidence in the company or the brand. If a company is unwilling to

[b] NAD Procedures at http://www.asrcreviews.org/asrc-procedures/.

participate, NAD will also refer the matter to the FTC with a request that the FTC review the advertising claims. NAD cannot legally compel the FTC to do so, but the FTC tends to support a self-regulatory process which holds advertisers to high standards of truth and accuracy. The FTC officials have publicly indicated on numerous occasions that the NAD referrals may be given priority by the FTC when they are received. Advertisers who initially refuse to participate before NAD often end up doing so after receiving a call from the FTC.

The advertiser has 3 weeks to respond to the complaint and to submit its substantiation for the claims that have been challenged. It is in this submission that the advertiser is expected to provide the evidence establishing a reasonable basis for its claims. The challenger then has 2 weeks to submit a written reply, and the advertiser finishes the briefing process with its final response 2 weeks later. The scheduling of submissions is somewhat flexible—extensions of these deadlines are often sought and received. After the briefing is completed, both parties meet with NAD separately—first the challenger, and then the advertiser. During these meetings, the NAD attorneys will ask the challenger and advertiser detailed questions about the substantiation submitted by the advertiser, any contrary evidence submitted by the challenger, and the parties' positions with respect to advertising claims that are made.

After meeting with the parties, NAD reviews the parties' submission and drafts a written decision that outlines the claims reviewed by the NAD, summarizes the parties' positions, and sets forth NAD's decision, including its findings and recommendations. If NAD believes there are deficiencies in the advertiser's substantiation either because the testing was not conducted properly or because it fails to substantiate the claims that are being made, it will recommend that the advertiser discontinue the challenged advertising claims or modify those claims in some way to better fit the substantiation. If it concludes that the substantiation is sufficient, its decision will say so. NAD's decisions are worded as recommendations, but most national advertisers agree to abide by these recommendations because if they do not, NAD will refer the matter to the FTC for investigation along with its decision indicating that the advertiser's substantiation is insufficient.

Because NAD attorneys review claim substantiation tests on a regular basis, they have an independent sense of how claims testing should be structured to support advertising claims. Whereas a court may need to hear testimony from experts about what is customary and reasonable for claims testing, NAD attorneys have some familiarity with claims testing procedures and they bring that experience to bear in resolving questions about the reliability and sufficiency of an advertiser's substantiation. As a result, NAD decisions tend to address more detailed aspects of claims testing than one is likely to find in a judicial decision. Additionally, NAD's decisions, published in *NAD/CARU Case Reports*, not only provide the outcomes of individual disputes but also an analysis of claim substantiation.[c] The *NAD/CARU Case Reports* library thus provides a source of claim substantiation guidance that can be relied upon by industry in developing new advertising claims.

THE NAD APPROACH

In determining whether an advertiser has a "reasonable basis" to support its claims, a key factor to be considered is the level of substantiation that experts in the field believe to be reasonable. Consequently, the use of consensus industry standards as claim support can be influential in the NAD's decision-making. In the case of consumer sensory testing, the NAD gives considerable weight to studies conducted in accordance with ASTM-1958 guidelines. Additionally, particular industries may utilize other consensus test standards to support claims, such as those developed for vacuum cleaner performance,[d] for removing stains from clothing, or for the whitening of teeth. However, not all industry test methods are formulated for purposes of making product comparisons. The NAD has noted that ASTM test methods are not necessarily meant to measure how a product will function under real-world conditions but rather often reflect an industry consensus on how to conduct product testing, how to test certain key product characteristics relevant to a product's application(s), and, thus ultimately, how to market the product.

It is important to note that conducting testing in accordance with a recognized industry standard, although helpful, does not guarantee a successful result at the NAD. On numerous occasions, the NAD has examined the evidence and found that the underlying comparative product testing was valid and well-conducted but nevertheless concluded that it

---

[c] NAD/CARU Case Reports archive at http://case-report.bbb.org/search/search.aspx? doctype = 1&casetype = 1.
[d] See, for example, for vacuum cleaners, ASTM F608, ASTM F2607, and IEC60312-1 5.2., for stain removal ASTM D4265, for teeth whitening, Vita Shade Guide.

was not sufficient to support *the claim that was actually communicated* to consumers. This tendency underscores the importance of taking into account not only the express claim (or claims) stated in the advertising but also the messages that may be implied.

Where a "sensory" claim is made, or a claim announcing a consumer preference for one product over that of a competitor based on consumer testing, ASTM-1958 carries great weight with NAD. For example, when NAD reviewed advertising for a diaper pail that claimed to be "Odor Free" and "more effective than the leading brand," NAD was asked to determine whether an objective industry standard test for quantifying gas transmission was sufficient for claim support. NAD determined that, contrary to the advertiser's position, when making a sensory claim, offering an objective test as substantiation is not more appropriate than ASTM E-1958 (Standard Guide for Sensory Claim Substantiation), "the latter of which covers reasonable practices for designing and implementing sensory tests and validates claims pertaining to the sensory characteristics of a product."[e]

In some challenges before NAD, both the challenging party and the advertiser submit evidence and NAD must, as courts sometimes do, engage in a "battle of experts" to determine which testing is more reliable. In one such case, Frito-Lay was challenged by Procter & Gamble, the maker of Pringles Potato Chips, for claiming that Frito-Lay Stax was America's taste test winner over Pringles and that "America prefers the taste of Lay's Stax over Pringles."

In that case, both parties submitted testing conducted in accordance with ASTM E-1958, although, in each case, both parties introduced certain variations. NAD concluded that the national taste test conducted by Frito-Lay provided a reasonable basis to support the taste preference claim, and that the test conducted by Procter & Gamble, which demonstrated "no difference," was not sufficiently powerful to overcome the reasonable basis established by Frito-Lay. Significantly, the case also underscored the fact that the NAD believes that the preference testing should include a clear "no preference" option to be asked of respondents.[f]

[e] Safety 1st (Dorel Juvenile Group, Inc.), Case Report #4223, NAD/CARU Case Reports (2004).

[f] Frito-Lay (Lay's Stax), Case Report # 4270, NAD/CARU Case Report (2004). (Currently, ASTM also recommends the inclusion of "No preference" option.)

There are a number of features of the NAD process that distinguish it from a court proceeding.

- There is no formal discovery at NAD. Each party gets to choose what evidence to submit to NAD in support of its position and how to present it.
- There are no counterclaims at NAD. If an advertiser has issues with the challenger's advertising, it can pursue a separate challenge, but that challenge will be heard separately on a separate timetable and often by a separate NAD attorney.
- NAD proceedings are confidential until the decision is issued, and the parties are limited by the NAD rules from using the NAD decision for publicity purposes. In contrast, a lawsuit is public from the minute the complaint is filed, and the party filing a complaint in court can use the filing itself to call into question a competitor's claims.
- NAD does not require the challenger to submit survey evidence to demonstrate that a challenged advertisement is communicating implied claims. Whether an advertisement is communicating implied claims is often a key issue in any false advertising dispute, and the ability for a challenger to prove its case based on arguments alone makes it much easier for it to prevail.
- Finally, and perhaps most importantly, NAD places the burden on advertisers to substantiate their advertising claims, including any implied claims. As such, it serves to police the FTC's mandate that advertisers possess substantiation for each claim that they make before making it. In court, the challenger must affirmatively demonstrate that the claim the advertiser is making is false.

NETWORK CHALLENGES

A number of the American broadcast networks also permit companies to file challenges to a competitor's advertisements that question the sufficiency of the advertiser's substantiation. NBC, ABC, and CBS have formal challenge processes, as does the Viacom family of cable channels. Fox Broadcasting does not currently have a formal challenge process.

Each network has its own written procedures, but the process for each is very similar to the NAD's. Both sides present their position in writing, network personnel discuss any questions they have with the parties separately, and then the network issues a written decision. If the network believes that an advertiser's substantiation is insufficient to support any message that is communicated by the commercial, either expressly or by implication, the network can stop airing the commercial or require the advertiser to modify the commercial if it wishes to continue airing it on the network.

Networks review only commercials that air on their own network, and their decisions are binding only with respect to those particular commercials.

LANHAM ACT LAWSUIT

Instead of filing a NAD or network challenge, a company who contends that its competitor's advertising is false can also file a lawsuit in court. Under the US Lanham Act, 15 U.S.C. § 1125, companies who are injured by another's false or misleading advertising may bring a civil lawsuit in a federal court:

> Any person who, on or in connection with any goods or services, … uses … any … false or misleading description of fact, or false or misleading representation of fact … in commercial advertising or promotion, [that] misrepresents the nature, characteristics, qualities, or geographic origin of his or her or another person's goods, services, or commercial activities, shall be liable in a civil action by any person who believes that he or she is or is likely to be damaged by such act.

Many states also provide companies injured by false advertising the ability to file a lawsuit under state law.

The remedies available in a Lanham Act lawsuit are much broader than the remedies available at NAD. A court can enjoin an advertisement, or an entire advertising campaign, within days or weeks of a lawsuit's filing if it determines based on preliminary submissions from the parties that the advertising campaign is making a false or misleading statement of fact that is causing injury to the plaintiff. If the injured party pursues the case through trial, a defendant found liable for false advertising may also have to pay the plaintiff all of the profits it has earned from its advertising campaign.

Although the remedies that are available in court are broader than the remedies that are available through NAD or a network challenge, the burden of proof is also higher. In court, unlike before NAD, a plaintiff who alleges that its competitor's advertisement is making an implied claim must generally submit a consumer perception survey to support its position.

Consumer perception surveys can be expensive to conduct, and will be subject to close scrutiny by the other side. The general practice in conducting such a test is to recruit a universe of consumers who were targets for the advertising (generally likely purchasers), show them the advertisement, and then ask them a series of unbiased questions about what messages are communicated by the advertisement. The questioning typically starts with open-ended questions such as "What was the main message of the advertisement" and then gradually progresses to closed-ended questioning about the specific messages communicated by the advertisements. These surveys were traditionally conducted as mall intercept studies, but they are increasingly conducted online unless there is some particular reason that the survey expert believes makes a mall intercept survey more appropriate.

Unlike before NAD, the plaintiff in a Lanham Act lawsuit must affirmatively prove that its competitor's claims are false. It cannot generally make its case by showing that the advertiser lacks substantiation or by arguing that the advertiser's substantiation is unreliable. There is an exception to this general rule when it comes to "establishment claims."[g]

Establishment Claims. When an advertisement says that "tests prove X," (or establish X) the plaintiff can meet its burden to prove that the claim is false not only by proving that X is false, but also by proving that the advertiser's testing does not, in fact, prove X. The plaintiff prove this by demonstrating that the referenced test does not meet industry standards, was unfair, or was otherwise unreliable. The plaintiff can also prove this by demonstrating the advertising in question misrepresents the test results. As a result, any testing referenced in an establishment claim will be subject to close scrutiny if the claim is ever challenged through a Lanham Act lawsuit.

Lawsuits in court tend to be more expensive and more time-consuming than the NAD or network challenges. But court actions also permit counterclaims; as a defendant, you may welcome the opportunity to go on the offense and file counterclaims against the plaintiff's

---

[g] {TA \/ "C.B. Fleet Co. v. SmithKline Beecham Consumer Healthcare, 131 F.3d 430 (4th Cir. 1997)" \s "C.B. Fleet Co. v. SmithKline Beecham Consumer Healthcare, 131 F.3d 430, 436 (4th Cir. 1997)" \c 1} C.B. Fleet Co. v. SmithKline Beecham Cons. Healthcare, 131 F.3d 430, 435 (4th Cir. 1997).

advertising. Depending on the strength of each side's position, leverage can be created for negotiating a settlement.

Lawsuits also permit discovery and, when the plaintiff seeks an injunction on an expedited basis at the outset of the lawsuit, the discovery can also be expedited. Discovery can be invasive, stressful, and highly disruptive. Parties to a lawsuit are entitled to request all documentation related to any testing that the advertiser has conducted to support the challenged advertising claims, any pilot or preliminary tests, and any related correspondence. They can also seek such documentation from third-party vendors who were involved in the test.

Individuals who are involved in the testing—including third-party vendors—may be called to answer questions under oath about the testing and their role in it. Prior to the deposition, opposing counsel will studiously review all the documents that have been produced related to the testing, and any pilot test; opposing counsel then has discretion to pose questions about every aspect of the test methodology, seeking to demonstrate that the test deviates from industry norms or was conducted in a way that biased the results in the advertiser's favor.

Church & Dwight, the maker of Arm & Hammer Super Scoop Cat Litter, sued Clorox for claiming that its Fresh Step cat litter is "better at eliminating odors than Arm & Hammer."[h] To support its claim, Clorox relied upon a "Jar Test" in which 11 trained panelists rated samples of cat waste alone (as a control) and then cat waste with the respective cat litter ingredients on a "0—15" scale. The results showed that the Fresh Step ingredient reduced odor 32% more than Arm & Hammer did. In determining whether to grant a preliminary injunction to stop the advertising, the court held a hearing in which both parties introduced expert testimony. According to Clorox's expert, the sensory evaluation method employed by Clorox had been reviewed in textbooks and peer-reviewed journals, and was taught in more than 40 universities in the United States. Church & Dwight introduced expert testimony asserting that, among other things, Jar Test's unrealistic conditions said nothing about how cat litter performs in circumstances relevant to a reasonable consumer. The court agreed that the results of the Jar Test were "not sufficiently reliable to permit one to conclude with reasonable certainty that they established the proposition for which they were cited in Clorox's commercial" and enjoined the advertising. The case ultimately settled.

[h] *Church & Dwight Co. Inc. v. Clorox Co.*, case number 11-cv-00092,S.D.N.Y. (2012).

Barring a settlement, all relevant evidence regarding the truth or falsity of the advertiser's claims will be presented to a judge or jury for a determination.

REGULATORY INVESTIGATIONS

The FTC, state attorneys general, and other government agencies also have the ability to challenge advertising that they consider to be inadequately substantiated or deceptive for any reason. These agencies have the ability to sue companies for false advertising with no warning. In general, although, after an investigation, they typically pursue a voluntary settlement with the advertiser before resorting to litigation.

Investigations can arise in a number of ways. In some cases, the agency or attorney general's office will contact the advertiser by letter and request that the advertiser voluntarily produce information concerning its advertising. In other cases, the agency will serve the advertiser with a subpoena that legally requires the production of documents. Like document requests in a Lanham Act lawsuit, a regulatory request or subpoena can require an advertiser to produce all documents and correspondence in its possession related to its claims testing. Regardless of the manner of contact, it is always a serious matter when a federal, state, or municipal agency contacts a company regarding its advertising.

Every investigation is different, but there are three general outcomes.

- The advertiser persuades the regulatory agency that its advertising is fully substantiated, not deceptive, or otherwise not worthy of the agency's resources to investigate.
- The advertiser enters into a voluntary agreement with the agency that requires it to make specific changes to its advertising and, in many cases, pay a fine.
- The regulatory agency files a lawsuit or other regulatory action against the advertiser, because the agency and advertiser cannot come to terms on the terms of a voluntary settlement agreement.

Agreements with government agencies can result in strict limitations on aspects of a company's advertising that remain in effect for years. Settlements often require companies to pay financial penalties to the regulatory agency and/or to consumers.

Although government agencies do not typically bring enforcement actions against comparative advertising, such actions are possible where there is perceived consumer harm. Also, the FTC has, in recent years, brought lawsuits against reputable companies for making claims that the FTC alleged to be false or unsupported. For example, in 2011, Reebok advertised shoes with muscle toning and strengthening benefits. It claimed that its shoes were "proven" to "work your hamstrings and calves up to 11% harder" and promised to "tone your butt up to 28% more than regular sneakers. Just by walking." The FTC alleged that these claims were not substantiated by competent reliable scientific testing.[i]

Faced with the FTC lawsuit, Reebok entered into a settlement which included 25 million dollars in consumer refunds.

CLASS ACTION LAWSUITS

Class action lawsuits by consumers pose another potential risk for advertisers. Most states have statutory prohibitions on deceptive advertising that permit consumers to file lawsuits if they have been injured by a deceptive advertising claim. In many cases, however, the injury experienced by any one consumer is too small to justify an expensive lawsuit. As a result, consumers typically file class action lawsuits on behalf of themselves and others similarly situated. Class actions can transform a lawsuit where a consumer seeks 5 dollars in damages for himself into one that seeks $50,000,000 on behalf of the class of 10,000,000 consumers similarly situated.

Class action lawsuits can be controversial. Many commentators believe that class action lawsuits benefit the attorneys who bring them more than the consumers they represent. It is nonetheless unquestionable that class action suits have an impact on advertising practices. Most recently, for example, many large food companies have been moving away from labeling their products as being "all natural" because of a flurry of class action lawsuits attacking the propriety of such claims. Because there is no clear government standard as to what constitutes an "all natural" product, the very filing of such a lawsuit by a class of consumers creates risk for a company; a court or jury simply might not agree with the definition of "all natural" that a company has been using internally to label its products.

---

[i] Federal Trade Commission, Reebok to pay $25 Million in Customer Refunds to Settle FTC Charges of Deceptive Advertising (September 28, 2011) at https://www.ftc.gov/news-events/press-releases/2011/09/reebok-pay-25-million-customer-refunds-settle-ftc-charges.

Class actions are like Lanham Act lawsuits insofar as discovery is available to the plaintiffs, and the burden of proof to demonstrate a claim is false is generally on the plaintiff. They differ in that an advertiser does not typically have counterclaims it can press against the plaintiff. The risk is thus asymmetrical.

Class action suits rarely go to trial, and many do not even result in discovery directed to the advertiser's substantiation for its claims. Instead, many are resolved at an early stage after the court determines whether it is appropriate for the case to proceed as a class action. If the court determines that the case does not warrant treatment as a class because, for example, the situations of the consumers in the proposed class are too different from each other, there is little incentive for the named plaintiff and its counsel to pursue the case. If, however, the court determines that the plaintiff has demonstrated a basis for proceeding as a class, then the defendant has a strong incentive to settle the case, which often occurs.

Class action settlements typically require advertisers to compensate consumers for the harm they allegedly experienced. The compensation can come in various forms, but in many cases, the settlement terms require the impacted class members to take affirmative action to receive reimbursement. Studies show that the percentage that does is very low.[1] Settlements can also require advertisers to change some aspect of their advertising. In almost all cases, they require compensation of the plaintiff's class action attorneys.

PLAYING DEFENSE CHECKLIST

☐ Evaluate at the outset whether the claim is one that is likely to be challenged by a competitor. Is the competitor litigious? Is the claim going to tout superiority as to a product attribute where the competitor claims to have an advantage? If the claim is likely to be controversial, plan for the worst.

☐ Make sure at the outset that there is a qualified person (other than counsel) who can explain and defend the testing procedures if necessary.

☐ Make sure you have a reasonable and fair rationale for any key decisions that you make with respect to your test methodology—particularly if they deviate from standard practice.

☐ Make sure that your claim substantiation evidence addresses not only the express statements made in the advertising but also any claims that are reasonably implied.

☐ Be careful of what you say in writing. If you have concerns, pick up the phone to communicate them to other relevant members of your team.

REFERENCE

1. Fisher D. Study shows consumer class-action lawyers earn millions, Clients Little. *Forbes* 2013.

CHAPTER 9

Into the Fray: Playing Offense

KEEP YOUR ANTENNAE UP

Nobody watches your advertisements more closely than your competitors and you may well do the same. A corporate watch program is advisable in any case to monitor potential infringements or abuse of your company's trademarks. Competitive intelligence gathering, known as "CI," is now a well-established corporate activity, supported by certifying institutions and codes of ethics in several countries. It is not to be confused with illegal industrial espionage. Some companies have departments or management positions responsible for CI gathering. Although it would be rare for companies to disclose their strategies or resources committed to CI, the evidence of their investment is revealed in the advertised CI management positions for which they recruit candidates.

The Internet has enhanced the power of corporate watch programs dramatically. Companies can employ services to receive instant alerts of every change that its competitors make to digital communications, including new advertisements, price changes, or product improvements. It is straightforward to incorporate a comparative advertising watch program into an already-existing CI strategy.

Each comparative advertisement by a competitor should be scrutinized for truth and fairness. Special attention should be paid to the consistency between words and visuals in televised or video comparative advertisements: precedents show that this type of inconsistency is a persistent source of misleading impressions.

BUSINESS-TO-BUSINESS RESPONSE

What are your options when you believe a competitor is making false or unsubstantiated claims, or misusing your trademark in a comparison? One option is to do nothing and just keep on competing in the marketplace. "As long as they spell our name right" is how some executives have expressed a belief that any advertising is good. Whether or not you subscribe to that belief, it is wise to stay on top of the options and

Ruth M. Corbin, (Editor): Practical Guide to Comparative Advertising
Into the Fray: Playing Offense, David Mallen and Thomas Jirgal, Principal authors.
DOI: https://doi.org/10.1016/B978-0-12-805471-0.00009-1. © 2019 Elsevier Inc. All rights reserved.

remedies available to you to address advertising claims that you believe are damaging your brand and company. The forums for initiating a complaint are the same as those that advertisers themselves were cautioned about in Chapter 8, Into the Fray: Playing Defense.

The first avenue of action to consider when encountering a comparative advertising program that targets your company is often a direct telephone call or letter by your corporation's lawyer to a corporate executive of the advertiser. This is more likely to be a productive option when you have a relationship and open channel of communication with your competitor. The executive team may be unaware of the provocation caused by the advertisement or indeed may have approved it without studying the justification in detail. If the advertisement is clearly false, or misleading, your competitor may be willing to consider voluntary modifications to the advertisement to avoid a dispute escalation.

If progress fails through an exchange of letters, then the cautions to defenders in Chapter 8, Into the Fray: Playing Defense, show exactly the avenues for plaintiffs to launch complaints. We list the pros and cons of each in turn.

SELF-REGULATORY CHALLENGE

In most countries with laws of fair competition and false advertising, an advertising industry association or professional accreditation associations have developed rules of conduct and processes for adjudicating allegations breaching them. Rather than taking a complaint through the costly and time-consuming court process, complainants have an option to avail themselves of the dispute resolution processes offered through self-regulatory organizations. The trade-off upsides and downsides you may want to consider are listed as follows:

Upsides, Depending on Country
- Less expensive than court.
- No counterclaims permitted.
- No requirement to submit survey evidence.
- Burden of proof is on the advertiser to prove its claims are true.
- Whether or not a complaint succeeds, some fault in the advertisement is likely to be identified, which may serve to constrain the advertiser's future aggressive comparative advertising.

Downsides, Depending on Country

- No award of money damages.
- Immediate injunctions not available; length of the decision-making process may fail to stop advertising campaigns of short duration that are unlikely to be repeated.
- Decisions are not legally binding (though companies found at fault are likely to comply).
- In case of labeling claims, some self-regulators, like the National Advertising Division (NAD), may permit the advertiser to exhaust existing products in the marketplace.
- Publicity surrounding filing of a complaint may not be permitted.

NETWORK CHALLENGES

Given the control of network corporations over their own content permissions and their own stake in protecting their network's reputation, a direct challenge before a broadcast network may be an attractive business option.

Upsides, Depending on Country

- Straightforward and efficient to implement.
- Networks have the ability to force an advertiser to modify its objectionable advertising on a faster schedule than NAD.

Downsides, Depending on Country

- Challenges are not available for all networks. In the United States, access to challenges is available through a limited number, including CBS, ABC, NBC, and Viacom cable channels. The readers should investigate those in their own jurisdictions and be aware that requirements may vary.
- A challenge would only pertain to television commercials airing on the network and not to other forms of advertising that may make the same claim.
- Some networks, like USA's NBC and Viacom, will not open a challenge if the challenger has filed a regulatory challenge through, for example, the NAD. Pursuing a challenge with those networks would require an advertiser to delay an industry regulatory complaint.

- Networks may reach different decisions for the same campaign. A network's decision only applies to commercials that air on that network. An advertiser can continue to run the same advertisement on other networks and national cable channels.

LITIGATION OPTIONS

Pursuing a litigation option entails very specific evidentiary requirements. If an action is started under America's Lanham Act, a plaintiff must establish the following:
- The defendant has made false or misleading statements of facts concerning his own product or another's.
- The statement actually or tends to deceive a substantial portion of the intended audience.
- The statement is material in that it will likely influence the deceived consumer's purchasing decisions.
- (In the United States) the advertisements were introduced into interstate commerce.
- There is some causal link between the challenged statements and harm to the plaintiff.[a]

For court actions in Europe, the preamble to the European directive is phrased as a set of grounds when comparative advertising is permitted rather than prohibited. The rules for being permissible identify, by implication, the grounds for offended competitors to take legal action:

"Comparative advertising shall, as far as the comparison is concerned, be permitted when the following conditions are met:
- it is not misleading
- it does not create confusion in the marketplace between the advertiser and a competitor or between the advertiser's trademarks, trade names, other distinguishing marks, goods or services and those of a competitor
- it does not discredit or denigrate the trademarks, trade names, other distinguishing marks, goods, services, activities, or circumstances of a competitor

---

[a] *See Grubbs v. Sheakley Group, Inc.*, 807 F.3d 785, 798 (6th Cir. 2015) and *Pernod Ricard USA, LLC v. Bacardi USA, Inc.*, 653 F.3d 241, 248 (3rd Cir. 2011).

- it does not take unfair advantage of the reputation of a trademark, trade name or other distinguishing marks of a competitor or of the designation of origin of competing products
- it does not present goods or services as imitations or replicas of goods or services bearing a protected trademark or trade name."[b]

Whatever the jurisdiction, a complaint through litigation channels is usually accompanied by a request for "damages" to be paid out in their monetary equivalent. The Lanham Act, for example, allows plaintiffs to claim lost profits, corrective advertising expenses to help regain loss in business, attorneys' fees, and punitive damages. However, standards of proof for recovering damages for false advertising are typically high and require plaintiffs to invest in establishing evidence. Typically, to be entitled to damages, plaintiffs must show that a material percentage of consumers were actually deceived or misled by the defendant's advertising and that such misleading or deception was a direct cause of injury to the plaintiff.

Punitive damages may be available for persistent willful falsity. A notably successful case for the plaintiff occurred in *U-Haul International Inc., v. Jartran, Inc.*, 793 F.2d 1034 (9th Cir. 1986). Jartran, the court found, had run a willful and malicious advertising campaign for more than a year, making false statements about its superiority over competitor U-Haul. A $40 million judgment against it included a requirement to run corrective advertising. The judgment was double the actual costs of damage and corrective action.

Litigation upsides and downsides for an offended corporation may be summarized as follows:

Upsides, Depending on Country

- Possibility of quick relief through preliminary injunction hearing.
- Availability of discovery.
- Availability of damages.
- Can potentially call competitor's claims into question by the mere filing of a suit.

Downsides, Depending on Country

- Expensive.
- Disruptive to business.

[b] Directive 2006/114/EC of the European Parliament and of the Council of 12 December 2006 Concerning Misleading and Comparative Advertising, Article 4.

- Can provoke counterclaims.
- Judges are unpredictable as to their interest level, ability to evaluate complicated advertising issues, and ability to hear a preliminary injunction motion on an expedited basis.

YOUR CHECKLIST FOR TAKING ACTION AGAINST AN ADVERTISER WHO STEPS OVER THE LINE

☐ Put into place a corporate watch program with specific attention to the consistency between words and visuals of advertisements in your corporate sector.

☐ On discovering an offending comparative advertisement, start by direct business communications to the advertiser.

☐ If the offending advertisement is televised, consider the pros and cons of an efficiently implemented complaint to the network.

☐ If more efficient options do not work, assess the trade-offs involved in an action through the regulator or courts. The first decision is whether the business is ready for a public confrontation. The budget required should be prudently estimated; action is frequently more costly than what people expect.

CHAPTER 10

Vive la Difference—Adapting Comparative Advertising to Different Countries

BUSINESS IS GLOBAL

Large corporations with offices in different countries frequently wish to show the same or similar advertising. At least two reasons account for this. One is global brand positioning. Famous world brands can become famous in part by repeating what they stand for in every country. Coca-Cola's memorable campaign featuring people of all ages and national backgrounds singing together "I'd like to teach the world to sing in perfect harmony" was an iconic representation of worldwide affection for the brand. Another reason that multinational countries wish to adapt existing advertising is cost. Advertising campaigns can cost in excess of a million dollars to design and produce. It is advantageous to capitalize on the investment by reusing as much content as possible—yet without compromising the likelihood of success in different international markets.

Still, there are many reasons that changes will be *factually* necessary from country to country, including the language of communication and different competitive products. Advertisers wonder, "what can I adapt from the comparative ad campaign that has already been designed for my home country?" Two criteria should be kept in mind when tweaking or adapting a comparative campaign designed originally for a different country market:

- Staying out of trouble with respect to differing laws and regulations of the respective countries.
- Tapping into meaningful cultural ideas that make comparative advertisements successful, and avoiding cultural offense or incongruity.

THE NAME IS THE SAME, BUT THE RULES DIFFER

"Comparative advertising" has been explicitly defined in the laws, directives, or regulations of several countries. Additional nuances arise when

Ruth M. Corbin, (Editor): Practical Guide to Comparative Advertising
Vive la Difference—Adapting Comparative Advertising to Different Countries, Dr. Ruth M. Corbin, Principal author.
DOI: https://doi.org/10.1016/B978-0-12-805471-0.00010-8. © 2019 Elsevier Inc. All rights reserved.

rules about comparative advertising are linked to the use of another company's trademarks.

Country laws and regulatory attitudes vary in their support. The United States is among the most constructively supportive countries of comparative advertising. Highlighting benefits to the consumer, America's Federal Trade Commission has produced a "Statement of Policy regarding Comparative Advertising" containing the following: "Comparative advertising, when truthful and non-deceptive, is a source of important information to consumers and assists them in making rational purchase decisions. Comparative advertising encourages product improvement and innovation, and can lead to lower prices in the marketplace."[a] Canada's laws and regulations that respect the use of trademarks in comparative advertising are similar to those of the United States.[1] Countries in the European Union have also more recently embraced the benefits of comparative advertising, with the European Directive 2006/114/EC explicitly noting that the comparative advertising directive applies to cases where competitors are named or *implied*.[b] The United Kingdom has interpreted the latter directive with explicit cautions against abuse of competitors' trademarks,[c] although it is otherwise well accustomed to accommodating comparative advertising. Legislation in Russia makes no mention of comparative advertising at all, but any form of negative comparison by an advertiser risks being found to be "unfair competition" or "fraudulent publicity" according to legal directives.[d] Indian law also gives no official definition to comparative advertising but counters abuse through its *Trademarks Act* and *Monopolies and Restrictive Trade Practices Act* to prohibit "unfair trade practices" and "disparagement." These examples pertaining to different countries are given to illustrate a single point: a multinational company launching a comparative advertisement in different countries must systematically check whether its use of a competitor's mark or implied identity will comply within each country's laws. Regulations may differ across countries with respect to where advertisements of particular

[a] FTC, 1979, at Paragraph c.

[b] The European Union's "Directive 2006/114/EC concerning misleading and comparative advertising."

[c] UK's "Business Protection from Misleading Marketing Regulations 2008," No. 1276, Part 1, Regulation 4, available online at http://www.legislation.gov.uk/uksi/2008/1276/regulation/4/made, accessed June 2016.

[d] As summary by a recognized legal analyst at http://www.lidings.com/eng/articles2?id = 6.

types can be placed, what times of day they may be advertised on radio or television, and what can be said—particularly with respect to regulated products like tobacco, alcohol, and pharmaceutical drugs. Local legal counsel should be retained.

An example of country-to-country inconsistency is found in the treatment of a commercial by SodaStream, manufacturer of a machine that turns ordinary water into carbonated water by injection of gaseous bubbles. Its commercial featured different people using their SodaStream machines. Each time one pressed down on the machine, a case of unidentified bottled soft drink spontaneously combusted in some other location in the country, with a voice-over saying "With SodaStream, you can save 1,000 bottles a year."[e]

Britain's TV advertising regulator, Clearcast, disallowed the advertisement on the grounds that it denigrated the soft drink industry: "The majority decided," said Clearcast's spokesperson, "that the ad could be seen to tell people not to go to supermarkets and buy soft drinks, instead to help save the environment by buying a SodaStream. We thought it was denigration of the bottled drinks market." But the same advertisement had already run in the United States, Sweden, Australia, and several other countries without regulatory interference. An appeal by SodaStream of the ban in Britain failed to reverse Clearcast's decision.

CULTURE-BASED ACCEPTANCE

Multi country campaigns are vulnerable to an equally important issue of cultural acceptance. The Pepsi-Cola company discovered that in a hard way, after significant setbacks in trying to export its Pepsi Challenge campaign. An advertisement featuring rap music icon M.C. Hammer was refused by all five major television stations in Tokyo, for its appearance of unfairness, despite the campaign later being found to be lawful. A Pepsi superiority campaign was similarly refused in Argentina, whose laws at the time included a prohibition against the use of a competitor's trademark as "unfair competition."[2]

Comparative advertising is less likely to be effective in cultures that value harmony and discourage confrontation, or those with "collective"

---

[e] http://www.campaignlive.co.uk/article/clearcast-forces-sodastream-pull-denigrating-ad/ 1161191#p36k6PEdqY7i2LmS.99; Clearcast's press release announcing decision and appeal at http://www.clearcast.co.uk/press/press-release-sodastream-ad/.

rather than "individualist" orientations to the society.[f] In collectivist cultures, Thailand being one where many studies have taken place, comparative advertising that stresses similarities rather than differences is more likely to be effective. Offense to cultural attitudes means damage to the advertiser's trademark value. Protecting that value goes beyond merely staying on the right side of the law. Even where the comparison itself is within the laws and norms of a country, other advertising elements need vetting for their cultural congruity and effectiveness. Comfort with new-age gender roles varies in different countries, for example.[3] The style of comparison, whether one or more competitors are explicitly named or merely implied, also varies in effectiveness in different countries.[4] As one expert in cross-cultural advertising has colorfully proposed:

> While the British aim for cuteness and are sometimes funny, Americans explore a lot of emotions like hunger, sex, fatherhood, and so forth. In advertising, the Japanese share the French attraction to allegories, showing the brand in context. Half-words are second nature in Great Britain, the country of understatement. Spain makes a specialty of unexpected demonstrations and visual unforgettables. German advertising assumes responsibility for being advertising. German ads seek to sell, they strive to convince. Norwegian advertising is characterized by crazy, random humor. In Asia, there is a humility and a humanity that give messages a very particular sensibility.[5]

Where there is no fundamental reason to expect cultural pushback in a country of interest to advertisers, academic studies encourage advertisers to be the first, where possible, to introduce comparative advertising in their product category. The novelty of comparative advertising among populations with little previous exposure to it appears to draw interest and engagement. For example, at a time when comparative advertisements were still a novelty in Korea, a head-to-head comparison between two desktop computer brands was found to be notably more effective in Korea than in the United States.[4]

All in all, comparative advertising by multinational companies requires investment in customization to laws and cultures of different countries.

---

[f] Korea is one such culture, as described by marketing text authors Wayne Hoyer and Deborah MacInnes in *Consumer Behavior (2008)*, Manson, OH: Cengage Learning, at page 135. Author Meghna Singh of the University of Delhi presents a broad-based analysis of comparative advertising effectiveness across different countries and cultures in "Comparative Advertising Effectiveness with Legal and Cross-culture Framework" (2014), *International Journal for Research in Management and Pharmacy*, Vol. 3, Issue 3.

It is fruitless to hope that it would involve nothing more than dubbing into different languages.

NEW-AGE IDEAS IN INTERNATIONAL MARKETING

Instead of framing a new entry strategy based on a country, some corporate innovators have thought first of how to characterize an untapped market internationally and chosen the country location thereafter. For example, in line with its broader social responsibility values, SC Johnson sought to promote its cleaning products to underdeveloped countries, with a view to supporting the global struggle against malaria.[g] With financial support from the Bill and Melinda Gates Foundation, it designed a family-share program for impoverished communities whose unsanitary living conditions gave rise to a high risk of malaria transmission. The family-share program enlists families as members of a club that can pool money together to buy cleaning and pest control products; the club benefits include refillable containers and loyalty points. Families end up not only sharing resources to purchase the products but also sharing their ideas and experience about keeping their homes safe from malaria-carrying mosquitoes.

The program was first piloted in Ghana. SC Johnson has since expanded its program of insect protection to families in need to other world locations. The company formulates and sends its products to appeal to local preferences. Promotion of its social responsibility programs has alluded to product superiority, without calling out competitors by name: "best possible value," "easier to use products," and "more affordable for more people." The company has received no significant pushback on these claims. While the program has not incorporated direct-to-consumer comparative advertising, it demonstrates the necessity and principles of adapting marketing strategies to diversified international markets.

YOUR INTERNATIONAL ADAPTATION CHECKLIST

☐   Famous brands maintain consistent positioning worldwide. The opportunity presents itself for an international company to communicate its desired competitive positioning in different countries.

---

[g] Reported by the company at http://www.scjohnson.com/en/commitment/socialresponsibility/protectingfamilies.aspx, last visited November 30, 2017.

Deciding to take hold of that opportunity is a matter of top-level strategy for the company.

☐ Engage local legal counsel in each country where the campaign is planned to advise on local laws and regulations and to vet copy.

☐ Pretest advertisements in each country for cultural compatibility.

☐ Avoid disparagement anywhere. Notwithstanding intercountry differences in nuances of permitted comparisons, almost all countries of interest to international advertisers prohibit disparagement of competitors' products.

REFERENCES

1. Pritchard B, Corbin R. Avoiding landmines. How to support comparative advertising claims. In: *Presentation to a conference sponsored by Advertising Standards Canada*; February 27, 2012. Available at: <http://corbinpartners.com/wp-content/uploads/2012/12/ASC-Presentation-Avoiding-Landmines-How-to-Support-Comparative-Advertising-Claims.pdf> [accessed June 2016].
2. White M. *A short course in international marketing blunders: mistakes made by companies that should have known better*. Brno, Czech Republic: Tribun EU; 2009.
3. Baxter SM, Kulcynski A, Ilicic J. Ads aimed at dads: exploring consumers' reactions towards advertising that confirms and challenges traditional gender role ideologies. *Int J Advert* 2016;**35**:970−82.
4. Jeon JO, Beatty SE. Comparative advertising effectiveness in different national cultures. *J Bus Res* 2002;**55**:907−13.
5. de Mooij M. *Consumer behavior and culture: consequences for global marketing and advertising*. Thousand Oaks, CA: Sage Publications; 2004. p. 216.

CHAPTER 11

Advertising Claims in Social Media

OVERVIEW

The marketing revolution is being tweeted.

Social media has revolutionized the way that we communicate and share information. It has also dramatically changed the way that companies market their products and services. Social media allows businesses and their brands to connect with customers in an engaging and creative way. It also provides marketers with unprecedented amounts of data, allowing brands to reach many customers with unprecedented efficiency. Social media connects consumers, creates communities, and changes the way that people process information and interact with one another.

Companies are increasingly using social media platforms (Facebook, Twitter, Instagram, Snapchat, and whatever is coming next), in creative ways, by way of launching targeted promotions, facilitating user-generated content (UGC), incentivizing influencers to talk about their products, and encouraging customers to share their experiences within their social networks. As a result, the face of advertising has been radically transformed. Every major advertiser has a social media marketing department and engages agencies and various advertising networks to maximize consumer engagement on social media. This rapid transformation of media has opened doors and created marketing opportunities which, only a decade ago, would have been unimaginable. It has also created uncertainty, anxiety about privacy and questions about the distinction between advertising and other forms of communications. The distinction is important, in part, because in many countries commercial speech is regulated differently than noncommercial speech. In the United States, for example, commercial speech does not have the same heightened protection under the First Amendment as other forms of speech. There remains considerable confusion over when and to what degree viral advertising, buzz marketing, paid bloggers, and other forms of social media campaigns are held to the same standard as other forms of advertising.

Ruth M. Corbin, (Editor): Practical Guide to Comparative Advertising
Advertising Claims in Social Media, David Mallen and Thomas Jirgal, Principal authors.
DOI: https://doi.org/10.1016/B978-0-12-805471-0.00011-X. © 2019 Elsevier Inc. All rights reserved.

This chapter addresses some of the rules and guidelines for advertising in social media and explores what social media means for comparative advertising and claim substantiation.

HOW SOCIAL MEDIA WORKS

Successful social media marketing campaigns take advantage of the unique features of the particular social media platform. Platforms such as Facebook, Twitter, YouTube, Instagram, Tumblr, Vine, and Snapchat each operate differently and offer different ways of engaging followers. Savvy marketers learn how to navigate these platforms to generate buzz, boost brand awareness, and provide avenues for consumers to talk about the product or the brand. Some social campaigns involve promotions using the social network to reach customers. Many of the most interesting and successful campaigns involve engaging consumers in a way that spurs consumers to talk about the product and share information with their friends and followers.

For example, in 2017, Airbnb, the online hospitality service, launched a celebrated social campaign called "Live there." The campaign utilized various social media channels to create and share the experience of living in different locations around the world. Airbnb partnered with professional photographers and users to create beautiful travel images on Instagram, a social media platform that lends itself to visual images. The campaign stirred conversation, generated buzz, and drove substantial traffic to Airbnb's own site—all indicators of success for the business.

Another example that illustrates ways that brands tie social media campaigns with traditional advertising was Procter & Gamble's 2017 campaign for Mr. Clean. Mr. Clean, the cleaning product, and its bald human mascot clad in white were the subject of a provocative television commercial which aired during the highly watched Super Bowl (where a 30-second spot can cost $5 million). In the weeks leading up to the Super Bowl, P&G launched a social media campaign on Facebook and Twitter, featuring teaser trailers to generate buzz about a commercial that promised to be funny and sexy. Within 1 minute of the Super Bowl commercial airing, Mr. Clean generated more than 11,000 mentions in social media. The video on YouTube also generated more than 17 million views.

The Mr. Clean commercial featured a sexy Mr. Clean dancing and gyrating his way to a clean house with an impressed wife looking on. The commercial closed with the tagline, "You Gotta Love a Man who

Cleans." But imagine, for a moment, a different cleaning product and a slightly different advertisement campaign featuring one or more of the following:

- The product being shown as outperforming (outcleaning) a competitive product in a side-by-side demonstration.
- Social media influencers being compensated to Tweet about the "unrivaled cleaning power" of the product leading up to the Super Bowl spot.
- Consumers being offered free entry into a sweepstakes if they post what they prefer about the product.
- Paid bloggers writing that plant surfactants make the product more environment-friendly than traditional cleaners.
- Consumers being given a $10 coupon or a free sample if they agree to post a review about the product.

Each of these hypothetical examples is likely to result in claims, express or implied, about the product. In the first instance, with a side-by-side demonstration, the advertiser is required to possess substantiation, typically consisting of a "reasonable basis" to support the claim being made (cleaning superiority on kitchen grease, soap scum, or whatever cleaning performance is being featured). However, in the other instances listed, the company is not, for the most part, controlling the content or what is being said. It is the social media consumers making the claims. Are these "claims" at all? When is an advertiser responsible for the claims made about a product by a consumer? Because social media advertising depends on interplay, a conversation among consumers and brand owners, there is a blurring of the lines between what is traditionally thought of as advertising and other kinds of content. This blurring of lines has created confusion and challenges for advertising in navigating social media.

WHAT DOES IT MEAN FOR COMPARATIVE ADVERTISING AND CLAIM SUBSTANTIATION?

In some respects, social media advertising is no different from any other advertising. The same rules apply for advertisers. Advertising must be truthful and nondeceptive. Objectively provable claims must be supported by a reasonable basis. Claims against competitors can be the subject of challenge, and advertisers are well advised to ensure that both the express and implied claims that are communicated are supported by a reasonable evidentiary basis.

In other respects, however, social media is radically different from other forms of media in its reach, rapidity, in the way consumers are involved, and in the way communications are perceived by consumers. It may be difficult to discern whether a particular statement or communication even constitutes advertising—this in turn poses a challenge because nearly all countries require advertising to be recognizable as advertising. In other words, consumers have the right to know they are being advertised to. Social media disrupts that expectation because brand messaging can take place by word-of-mouth, where consumers themselves are not passive receptacles for advertising messages. They are conduits of the message—part of the advertising ecosystem.

Another complication arises when advertisers lose control of the trajectory of the messaging. Because advertisers are responsible for not only express claims but implied claims as well, the rapidly evolving context of any given social media conversation makes it very difficult to know whether certain messages taken up on the social media platform were "implied" by the original message, and even more difficult to measure the extent of misperception or its impact. An advertiser may communicate a certain performance benefit for its product, like "stronger enamel." But once consumers take part in the conversation, the message can change and give rise to other exchanges like an inferred message of "puts an end to cavities forever." One misguided individual can start a barrage of misimpressions, repeated from one person to another. The original advertisers thus run the risk of losing control of the messaging and of course of the conversation among a growing audience.

WHEN CONSUMERS SPEAK: THE CHALLENGE OF USER-GENERATED CONTENT

It is perhaps a small irony that the consumer engagement benefit of social media carries with it the bulk of the risks to the advertiser. Through social media, brands can engage consumers directly, leverage their voices, and spur the creation of user-generated content (UGC) in the service of the brand. This can result in some novel competitive comparisons but also some unintended consequences. True enough, when consumers talk about product performance, which products they prefer, and what they like and don't like, that is not advertising; it's just a conversation. But what happens when companies are seen to *encourage* consumers or to incentivize consumers to talk about the company's products and those of their competitors?

That question was put to the test. In 2006, Quiznos, the sandwich chain, sponsored a contest inviting consumers to create and upload videos comparing its cheesesteak sandwich to a Subway cheesesteak sandwich. As part of the instructions, consumers were told that the video should demonstrate "why you think Quiznos is better" and were asked to upload their videos to www.meatnomeat.com. In response to the invitation, consumers created and uploaded videos, many of which lampooned and disparaged Subway and claimed that in comparison to Quiznos, the Subway sandwich had less meat and meat of inferior taste and quality.

Although many of the videos were intended to be funny, Subway was not laughing. It sued Quiznos in Federal Court for false advertising under the Lanham Act. Quiznos asked the court to dismiss the lawsuit arguing that it was immune from liability under the Communication Decency Act, a law which treats the provider of an interactive computer service as a publisher (as opposed to an advertiser) and therefore immune from liability for objectionable content. The court denied Quiznos' motion and the parties eventually settled. However, the case left open the possibility of an advertiser being held liable for false claims made as part of a UGC promotion. In other words, the advertiser may be responsible for false impressions or claims communicated by consumers within the context of the promotion.[a]

The US self-regulatory organization, National Advertising Division (NAD) has also wrestled with comparative claims made as part of UGC in social media channels. In 2011, Nestle challenged LALA USA, Inc., the maker of dairy products, over a UGC promotion in which consumers were asked to create and share vignettes that explore "where non-dairy creamer comes from." The vignettes claimed that non-dairy creamers contained ingredients also found in paint, glue, shampoo, and shaving cream, and some non-dairy creamers were actually depicted as a replacement for glue or paint. Consumers were invited to post these video vignettes on their own sites and to share them with their friends. During the course of NAD's review, the advertiser said it would permanently

---

[a] Doctor's Associates, Inc. v. QIP Holder LLC, 2010 WL 669870 (D. Conn.).

discontinue the posting and sharing of the challenged vignettes.[b] NAD's decision however noted that by encouraging, facilitating, or sharing the creation of content that was false or unsupported, an advertiser may be responsible for any unsupported or misleading claims.

RULES FOR SOCIAL MEDIA

Because technology moves faster than developments in the law, social media is often considered the "wild west" of advertising—a medium where anything goes. There is, however, a body of law and guidance that governs how companies can advertise using social media. The use of social media by companies implicates a host of legal areas ranging from Employment Law, Intellectual Property, Privacy, to Cybersecurity. For purposes of this volume, we will focus on the rules and guidance that relate to advertising practices themselves, including those that govern comparative advertising.

One issue that frequently arises when statements about products and companies are made or implied by consumers is whether those statements are considered endorsements or testimonials made on behalf of an advertiser. A key source of guidance in the United States, therefore, comes from the Federal Trade Commission's (FTC) revised *Guides Concerning the Use of Endorsements and Testimonials in Advertising* ("FTC Endorsement Guides").[c] In revising the guide, the FTC reaffirmed some general principles of advertising law. For example:

- an advertiser cannot use a consumer's testimonial to make an otherwise unsubstantiated claim;
- a consumer's comments about a product and his or her experience and results with the product must be independently substantiated;
- if the customer providing a testimonial has experienced atypical results, the advertiser must disclose the average expected results (providing a disclosure indicating "results not typical" is not sufficient).

Because the 2009 FTC endorsement guides are intended to address advertising generally and because the release preceded the latest explosion in social media marketing, companies continue to face questions about

[b] LaLa USA (La Crème), Case Report # 5359, NAD/CARU Case Reports (2011).
[c] 16 C.F.R. § 255 (2009)), available at https://www.ftc.gov/sites/default/files/attachments/press-releases/ftc-publishes-final-guides-governing-endorsements-testimonials/091005revisedendorsementguides.pdf.

when an engagement with a consumer constitutes an endorsement or testimonial, and about how to properly disclose "material connections" between the consumer/endorser and the advertiser. As a result, the FTC issued additional guidance on its FAQ page, titled *The FTC Endorsement Guides: What People are Asking* (2017).[d]

The FTC provides the following guidance with respect to advertisers who use social media:

- *Material connections must be disclosed:* If a relationship or "material connection" would impact the credibility of the claim or message, that connection must be disclosed.
- *A picture can be an endorsement:* Simply posting a picture of a product in social media, such as on Pinterest, or a video of you using it could convey that you like and approve of the product. A picture can also give rise to an implied claim, requiring substantiation.
- *Using a "like" button or sharing a link can be an endorsement:* If there is an inducement or a consumer "likes" "pins," or "shares" a picture or product message as part of a paid campaign, that may constitute an endorsement.
- *A consumer who reviews product and receives an incentive is an endorser:* The "material connection" must be disclosed in a clear and conspicuous manner.
- *Contests and sweepstakes that have been incentivized must be disclosed:* If a post about a product is made as part of a contest, the post should disclose this form of material connection (e.g., #sweeps).

A consistent theme emerging from the FTC guidance is the notion that consumers have the right to know that they are being advertised to. If that is not clear, the advertisement may be viewed as containing an implied claim that "this is not advertising." If there is a paid relationship, that claim or implied claim would be misleading.

By way of example, suppose a consumer (or celebrity) makes an endorsing statement in social media that "no skin cream works better on my skin." If an incentive has been provided, either by way of payment or some other "material connection" between the advertiser and the consumer, it may be presumed by a regulator that existence of such incentive (if social media users were aware of it) would affect the credibility of the consumer's statement. That presumption explains why social media users

---

[d] Available at https://www.ftc.gov/tips-advice/business-center/guidance/ftcs-endorsement-guides-what-people-are-asking.

should be made aware of testimonial incentives (the "material connection") through clear and adequate disclosure. Furthermore, in the event of such "material connection," an advertiser may also be responsible for having substantiation in hand for the endorser's claim.

For advertisers engaging in social media campaigns, an important first step is to identify any "material connection" that must be disclosed. This needs to be done even if a company does not know in advance exactly what consumers with a material connection will write. The FTC, in its *Dot.Com Disclosures*, offers important guidance for advertisers concerned with how to disclose material connections in digital media and social media platforms.[e]

Advertisers may find themselves wrestling with the challenge of making disclosures in the limited confines of social media platforms. However, if an advertiser later finds itself mired in a dispute about deceptive communications, including failure to disclose a "material connection," it is no defense to say that there was insufficient space. According to the FTC if the disclosure cannot be made, the claim should not be made. Advertisers have therefore resorted to creative means of disclosure including the use of hash tags (#) on Twitter-type platforms, where the 140-character limit is particularly restrictive. Here, FTC provides guidance.

In the example below, a celebrity tweets about her success with a weight loss product.

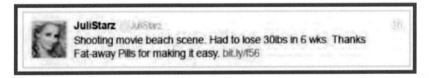

According to the FTC, the tweet is misleading because it does not communicate that she is a paid endorser ("Thanks Fat-away Pills" does not sufficiently disclose the relationship). It also fails on grounds of claim substantiation because the weight loss results, 30 lbs in 6 weeks, are not

[e] FTC, *How to Make Effective Disclosures in Digital Advertising* (March 2013) (*FTC DoTCoM Disclosures*, available at https://www.ftc.gov/sites/default/files/attachments/press-releases/ftc-staff-revises-online-advertising-disclosure-guidelines/130312dotcomdisclosures.pdf).

typical or supported by independent substantiation. The following example, according to the FTC, complies with the law:

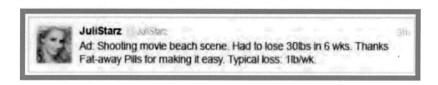

The Tweet begins by identifying that it is an advertisement and it also indicates the amount of weight loss that has been shown to be typical with this weight loss product (1 lb/week).

ENFORCEMENT AND THE RISE OF INFLUENCERS

A powerful means of advertising on social media is by means of paid "Influencers." Influencers may be celebrities themselves or consumers who have accumulated large followings through their social media networks. As a result of their followings, they can be effective and persuasive communicators of brand messaging. When they are paid to promote products, they are "endorsers," according to the FTC, and the fact that they are being compensated, or that they have a "material connection" to the brand, is a fact that must be disclosed.

FTC has taken action when disclosure of material connection has not been made or has been inadequate. In 2015, the FTC charged Machinima, Inc., a California-based online entertainment network, with engaging in deceptive advertising by paying "influencers" to post YouTube videos endorsing Microsoft's Xbox One system and several individual games.[f] According to the FTC, the influencers paid by Machinima, Inc. failed to adequately disclose that they were being paid for their seemingly objective opinions. The case resulted in a settlement prohibiting Machinima from engaging in such conduct in the future. The company is required to ensure its influencers clearly disclose when they have been compensated in exchange for their endorsements.

[f] https://www.ftc.gov/news-events/press-releases/2015/09/xbox-one-promoter-settles-ftc-charges-it-deceived-consumers.

What is notable about the FTC settlement with Machinima is that the FTC did not pursue enforcement action against Microsoft, the advertiser of Xbox One, the product at issue, or its advertising agency. The FTC noted that these companies had robust compliance programs and that the misleading claims to consumers were made in spite of their efforts and not because of them. This case illustrates the importance of companies' putting systems in place for training employees and implementing programs to monitor social media communications, to ensure compliance with essential disclosures.

The FTC reached a similar result when it alleged that Warner Brothers failed to adequately disclose that it paid online influencers to post videos promoting its video game, *Middle Earth: Shadow of Mordor* (https://www.ftc.gov/news-events/press-releases/2016/07/warner-bros-settles-ftc-charges-it-failed-adequately-disclose-it). According to the FTC, these videos were designed to generate buzz within the gaming community for the new release of a fantasy role-playing game loosely based on *The Hobbit* and the *Lord of the Rings* trilogy. The FTC alleged that Warner Brothers hired online influencers to develop sponsored gameplay videos and post them on YouTube and to promote the videos on Twitter and Facebook. This activity by Warner Brothers generated millions of views. PewDiePie's sponsored video alone was viewed more than 3.7 million times. Significantly, Warner Brothers required the influencers to promote the game in a positive way and not to disclose any bugs or glitches they found. The FTC further alleged that Warner Brothers failed to instruct the influencers to include sponsorship disclosures clearly and conspicuously in the video itself where consumers were likely to see or hear them.

When advertisers work with influencers, they must be aware that failure to disclose material connections can leave them liable for not just claims made by influencers on their behalf but for any resulting misleading of consumers.

CASE STUDIES

What do these challenges mean for comparative advertising in social media? Although the FTC and other government agencies rarely intervene in cases of comparative claims, the self-regulatory organizations have been active. In the United States, NAD provides the forum where most disputes relating to social media are resolved. NAD has been called upon to address and resolve a number of interesting issues relating to an advertiser's obligations when using claims in social media, including comparative advertising claims.

One interesting case from 2011 involved a #1 claim or what the advertiser's competitor alleged was a #1 claim. By way of background, it should first be observed that "hashtags" are a central feature of Twitter that Twitter claims to have originated and that have been adopted by other social media platforms. Hashtags employ the use of the "#" pound sign with a word or phrase to mark a person, idea, or concept within a tweet that may have broader relevance (e.g., #adclaim). Hashtags serve as a tool for searching for other related tweets that use the same hashtag, and also a means for Twitter to identify trending topics.

The question NAD confronted in *Bridgestone Golf, Inc. (Golf Ball Fitting)*, Case #5357, NAD/CARU Case Reports (2011) was whether a hashtag might also be a claim and, if so, whether the "#" pound sign itself should be considered part of the claim. The advertiser in question, Bridgestone Golf, had reserved the hashtag "#1BallFitter" for its Twitter account, and its competitor argued that consumers would perceive this hashtag as a claim that Bridgestone was "the number one ball fitter," which was a claim the competitor contested.

After evaluating the arguments for and against, NAD agreed with the competitor that a #1 claim was being communicated, or at least that a reasonable consumer may perceive the hashtag as a #1 claim in this context—that is, that the advertiser was the number one ball fitter. Among other things, NAD pointed out the clear intention of the advertiser to communicate a "#1" position, because the hashtag "1BallFitter" did not really make sense without the number sign. Also key to its finding was the fact that the advertiser was claiming elsewhere to be the #1 Ball Fitter. Given these facts, NAD's conclusion was hardly controversial; but the case serves as an example of how social media can be used to communicate an advertiser's message in novel ways.

Another NAD social media case, that received a large amount of attention in the advertising community, involved the use of "likes" from Facebook for advertising purposes. In *Coastal Contacts, Inc. (Coastal Contacts, Inc. Advertising)*, Case #5387, NAD/CARU Case Reports (2011), the challenger took issue with "like-gated" promotions—a practice that used to be permitted on Facebook, but no longer is. "Likes" are a key feature of Facebook, and individuals who participate on Facebook often strive to obtain as many "likes" as possible for their social media posts. The same is true with advertisers. Companies with Facebook pages generally strive to have as many Facebook users "like" their pages as possible. It helps them to distinguish themselves in the marketplace.

Facebook functionality previously permitted companies to run promotions to incentivize "likes" for their company's pages. The advertiser in question, Coastal Contacts, ran such a promotion where it offered Facebook users a so-called free pair of glasses if they "liked" the Coastal Contacts Facebook page. The challenger raised a number of issues with the advertiser's promotion, including the manner in which it only disclosed the conditions for obtaining the so-called free pair of glasses once users "liked" the Coastal Contacts page. NAD reviewed the challenged promotion as it appeared on Facebook and agreed that conditions on the advertisers' "free" offer should be communicated to consumers at the outset of the offer before the consumers were required to take any action, such as liking the page. This conclusion was consistent with the high standards that the FTC and NAD generally impose on the use of the word "free" in advertising, which require that conditions for such offers be disclosed even more prominently than is normally the case.

A more interesting question was whether the "likes" that Coastal Contacts obtained through this promotion were essentially tainted and should be removed on the grounds that they conveyed a message that 250,000 Facebook users (or whatever the number actually was) genuinely "liked" Coastal Contacts. The challenger argued that the total number of Facebook likes featured on a company's Facebook page—for example, 250,000 likes—was essentially a claim that 250,000 Facebook consumers legitimately liked the company. The challenger further argued that this was problematic in Coastal Contacts case because it obtained some number of these likes through a promotion that the advertiser contended was misleading.

NAD disagreed with the challenger's main premise that consumers would view a company's "likes" as all deriving from a general affinity for the company. NAD reasoned that like-gated promotions were permitted by Facebook, and that Facebook users, who were the primary audience for any claim communicated by a company's total number of likes, would understand that they "like" pages for a variety of reasons, including in response to promotions like the ones run by Coastal Contacts. As a result, NAD did not recommend that Coastal Contacts remove the likes obtained through this promotion.

Interestingly, Facebook itself seemed to recognize that the use of like-gated promotions was serving to undercut the value of "likes" for companies. A few years after NAD's decision, it ended up prohibiting such promotions through revised terms of use. According to an *Adweek* article, the move represented an effort by Facebook to make sure that "[w]hen you like a Facebook page, . . . you really like that page."[1]

A third NAD case[g] that dealt with multiple claims that a company made based on the same taste preference taste—some in traditional media and some in social media—presents a noteworthy example of what an advertiser can do with a well-supported substantiation study. The company in question, General Mills, was reintroducing its Greek Yogurt into a market that was rapidly changing and crowded with competitors. Looking to make a splash, it commissioned a third-party taste test to support an advertising campaign. The target for the taste test was carefully chosen by the company's internal marketing team: blueberry yogurt from Chobani, which was the leading flavor from the leading Greek yogurt maker. Outside counsel weighed in, the test was fielded, and the results were positive: on a nearly 2:1 basis, consumers preferred the taste of General Mills' Yoplait Greek yogurt to Chobani's. It was a home-run result.

General Mills advertised its strong taste preference results on television, in print, and online. It created shelf talkers to inform customers of the test results at the point of purchase. And it encouraged consumers to conduct their own "taste offs" between Greek yogurt from Yoplait and Chobani and tweet the results with #TasteOff hashtag. General Mills then took a curated selection of the hashtagged tweets and used them to populate a Yoplait Greek Yogurt Taste Off Tumblr page with the results. It was a multiplatform campaign that all centered on a single taste test, a test based on a single yogurt flavor.

Chobani immediately challenged the campaign at NAD. Within a week of seeing the first advertisement, Chobani filed a challenge at NAD to General Mill's entire advertising campaign. The primary basis of the challenge was the taste test, which Chobani said was unreliable due to the manner in which it was conducted. In particular, Chobani objected to the comparison of its fruit-on-the-bottom yogurt to Yoplait's preblended variety. Chobani argued that it was an unfair comparison and that the very manner in which the yogurt was presented to consumers in the taste test unfairly biased the results in General Mills' favor. General Mills defended the test to NAD and prevailed. NAD concluded that the test

[g] General Mills, Inc. (Yoplait) Case #5715, NAD/CARU Case Reports (2014).

was conducted appropriately and that its results were reliable. NAD also concluded that General Mills' television advertisement was appropriately limited to the results of the taste test—blueberry Greek yogurt—and did not communicate a broader line claim that Yoplait Greek yogurt generally tasted better than Chobani Greek yogurt.

NAD then reviewed the social media campaign. The first question that NAD addressed was whether the tweets tweeted by participants in the campaign who conducted their own taste test at General Mills' request were claims for which General Mills had to provide substantiation. Should NAD have found General Mills responsible for the substantiation of consumers' own tweeted words on the social media site, it would have created a challenge, since some of these tweets were not as carefully written as the Yoplait television commercial, which had of course been subject to legal review before airing. Survey participants had used broader language in their tweets that could have been perceived as broader claims than had been supported by the test, such as "I picked Yoplait Greek," or "Yoplait Greek Pineapple all the way." There was no practical way, argued General Mills, for it to review participants' tweets ahead of time, as that would have defeated the whole purpose of the campaign.

NAD concluded that consumer tweets, on their own, were not a problem. As it did in the *Coastal Connections* Facebook case, NAD concluded that within the Twitter ecosphere, other Twitter participants were likely to understand that the tweets were individual statements of opinion and not claims that required claim support.

The Yoplait Tumblr page was a different story. This page aggregated various different Twitter comments that included individual #TasteOff results like the ones referenced above (i.e., "I picked Yoplait Greek," "Yoplait Greek Pineapple all the way," etc.) and presented them on a page that also included the claim: "The #TasteOff is on! Nearly 2 out of 3 people agree—Yoplait Greek Blueberry Tastes Better than Chobani Blueberry Fruit on the Bottom." The question for NAD was whether consumers would view the combination of these more broadly worded tweets differently when they were presented in a different context. NAD concluded that this difference in presentation was significant and that consumers might, in this context, view the combination of messages— some approved, and some social—as an overall claim that Yoplait Greek yogurt tastes better than Chobani Greek yogurt overall. As a result, NAD recommended that General Mills makes some changes to the Tumblr page to help avoid such a broad message.

These NAD examples add to the evidence of challenges that can arise from social media campaigns, challenges that are not always easy to anticipate. They illustrate how traditional rules of comparative advertising are being applied in the rapidly evolving world of social media. And most usefully for the future, the cases provide guidance on evolving regulatory interpretations.

YOUR SOCIAL MEDIA ADVERTISING CHECKLIST

☐ As a matter of contemporary business strategy, every plan to launch a comparative advertising should consider an accompanying social media component.

☐ Stay alert to the risks of soliciting consumers to participate in online discussions of your product's superiority; such solicitation activities may be subject to regulations affecting comparative advertising through traditional media. Regulations concerning endorsements are also pertinent. The extra challenge for marketers is that there is little control over where consumers take the conversation on social media.

☐ Endorsers that are paid or otherwise associated by business interest to the advertiser must be disclosed.

☐ Administrators of advertising laws and regulations are continuing to be presented with novel cases that in turn are shaping the evolving law. Legal counsel should be consulted each time a new creative marketing idea is proposed for social media.

REFERENCE

1. Lafferty, J. *Facebook is demolishing the like gate.* Adweek.com; August 8, 2014. Available at: < http://www.adweek.com/digital/facebook-is-demolishing-the-like-gate/#/ > .

CHAPTER 12

Summary and Handy Checklist

Dare to Compare has been prepared as a guidebook for corporate market-ers, advertising professionals, and the lawyers who advise them. It may be helpful in training new staff. This chapter recaps highlights and assembles the checklists from each chapter. Once the content of the book has been reviewed, this chapter provides a memory aid for all the topics that are needed to keep in mind, under time pressure of a comparative advertising launch.

COMPARATIVE ADVERTISING: LOOK BEFORE YOU LEAP

The choice to be a "comparative advertiser" has implications much beyond the marketing department. Customer perception, employee morale, industry relationships, legal position, and financial risk/reward are all implicated in the choice. Once committed to a comparative campaign, a corporation will set itself on the trajectory of steps enumerated here:

☐ First, be aware of legal and regulatory restrictions for what can be claimed in a comparative advertisement. Chapter 2, Governing Laws, Network Standards, and Industry Self-Regulation, provides a brief education on that topic.

☐ Next, consider what claim of positioning against a competitor you want to make. The claim will likely complement the positioning you seek for your product or company overall. Chapter 3, What's the Name of the Claim, offers a catalog of claims from which to choose.

☐ Claims must be explicitly supported by evidence. Remember that you are responsible not only for the express claims but also for claims that may be implied to reasonable consumers. Chapter 4, Foundations of Test Design, describes the criteria that your evidence must address to withstand scrutiny before a regulator or judge.

☐ Comparative claims inevitably involve quantitative words like "equal to," "more," "majority," or "most preferred." Proving your claim will come down to statistics of some kind. How big does your statistic

Ruth M. Corbin, (Editor): Practical Guide to Comparative Advertising
Summary and Handy Checklist, Ruth M. Corbin, Principal author.
DOI: https://doi.org/10.1016/B978-0-12-805471-0.00012-1. © 2019 Elsevier Inc. All rights reserved.

have to be to qualify? The answer is less obvious than you think. See Chapter 5, Statistical Support—How Much Is Enough?

☐ With the preparatory steps above, a company is ready to add the magic of a brilliantly designed advertisement. There is no chapter in this book on how to summon up the magic. But once designed, a comparative advertisement needs to be fine-tuned. It may need, for example, qualifiers, explanatory notes, or disclaimers to avoid misleading its audience. It may need occasional tweaking to avoid obsolescence, particularly if your competitor makes changes to the very product being compared in your advertisement. Chapter 6, Know Your Limits: Claims Have Boundaries, reminds you about the fine-tuning issues.

☐ Anticipate the unexpected. Even when everything looks fine to the corporate team, a final disaster check is in order before launch. Chapter 7, An Ounce of Prevention: Troubleshoot Your Claim Before Launch, advises on those final steps to minimize the chance of a market fiasco once the advertisement is launched.

☐ Comparative advertisements provoke competitors. Once the advertisement is launched, marketers and their legal advisors should reserve time and money for addressing a possible legal challenge by the targeted competitor. See Chapter 8: Into the Fray: Playing Defense, on how to prepare.

☐ What if your company is the targeted competitor, rather than the protagonist, in a comparative advertising campaign? Chapter 9, Into the Fray: Playing Offense, provides advice on reacting to such a campaign.

☐ Corporations doing business in more than one country will find useful the international perspectives added in Chapter 10, *Vive la Difference—Adapting Comparative Advertising to Different Countries*.

☐ Advertising through social media is subject to all of the same principles summarized above. However, it brings novel problems. Some of the problems to anticipate when including social media in your comparative advertising campaign are highlighted in Chapter 11, Advertising Claims in Social Media.

LAWS, REGULATIONS, AND COMPLAINTS

Once a company sets out on a path of comparative advertising, it needs to give early attention to its obligations under applicable laws and regulations in its own country and in other countries in which it hopes to extend its campaign. Self-regulatory professional and industry organizations may add another layer of rules of conduct.

☐ Readers should compile the relevant (1) laws, (2) regulations, and (3) media rules for their own jurisdictions.

☐ From the early stages of formulating a comparative claim for future advertising, advertisers should obtain legal advice on the requirements to substantiate it and any precedents in law or regulation that should be taken into account.

☐ Even when rules are scrupulously followed, advertisers should anticipate competitive push-back in one form or another and have a plan—before the advertisement is launched—to respond accordingly.

☐ International advertisers need to revisit each of the summary steps above for each country in which they intend to extend their campaign. Ensuring locally relevant substantiation for a competitive claim will always be a good investment.

WHAT'S THE NAME OF CLAIM—A USEFUL CATALOG

The catalog of claims includes Pinnacle Number One Claims, General Superlative Claims, Targeted Superiority Claims, Explicit or Implied, Parity Claims, Customer Preference Claims, Testimonials, Sensory-based Claims, and Factual or Technical Claims. The list is subject to variations but is intended to give a structure for thinking about how your company wants to position or communicate its competitive strength.

☐ Decide on the competitive benefit you want to promote.

☐ Choose from among the options for expressing it. Choose one or more options for testing that you anticipate will be supported.

☐ Decide on whether you will identify a competitor by name or not.

☐ Seek valid and reliable evidence of the truth of the claim or claims being tested, evidence that could withstand scrutiny of an expert hired by your competitor, if an objection is later raised. The evidence should be tailored to the wording of the claim—as close to the "exact" wording as reasonable.

FOUNDATIONS OF TEST DESIGN

Decisions must be made and records should be kept for each of the following, where applicable:

- [ ] Decision on what specific claim is to be tested
- [ ] Choice of experimental design that serves the test of the claim, exactly as it is worded
- [ ] Methodology, in-person (central location or at-home), telephone, or Internet
- [ ] Selection criteria of the target population
- [ ] Samples sizes for each subgroup relevant to the experimental design
- [ ] The process for recruitment, screening, and qualification
- [ ] Test locations and dates
- [ ] Questionnaire
- [ ] Data collection process
- [ ] For in-person testing, interview/test staff instructions, including instructions on product handling, preparation, and serving
- [ ] Product procurement and storage
- [ ] Product preparation protocols
- [ ] Participant instructions
- [ ] Coding instructions for open-ended responses
- [ ] Data tables
- [ ] Assembly of raw data in readable form
- [ ] Data analysis and covering report

STATISTICAL SUPPORT—HOW MUCH IS ENOUGH

The use of statistics removes ambiguity. Once the technical experts confirm the legitimacy of how statistics are used, based on sufficient sample sizes and quality controls, then statistical statements can be turned into readily understood consumer claims Your readiness to incorporate the industry's major guidelines for support can be checked from the following list:

- [ ] Decide on the claim you want to make.
- [ ] Set up the wording for the hypotheses that will determine the test design.
- [ ] Get input at this stage from the advising lawyer, regarding any issues in principle with the wording of the claim. If issues arise, either modify the claim or ensure that the issues are covered off in the research design.
- [ ] Determine sample size, including subsample sizes, based on guidelines you will use as your authority.
- [ ] Recruit a sample using rigorous randomization methods and ensure that consume qualifications are matched to the audience for the claim.

□   Determine the quantum necessary for the claim to be supported.

□   If support is obtained, assess every possible reason why an opponent might disagree. Decide on whether the risks are tolerable.

□   Document definitions and assumptions, for production in the report.

CLAIMS HAVE BOUNDARIES. KNOW YOUR LIMITS

Even well-supported claims have their limits: limits on what can be said, what must be said, and how long the claim is available to be made. If a competitor acts quickly to modify its product, the investment in the advertising program may have limited time value. This should be antici-pated for fast action if necessary. Sometimes, of course, a short window of time after a product is launched—even if the advertisement is ordered off the airwaves—is enough for a company to make the impact it needs.

□   Ensure that the wording of the comparative claim matches the test that supports it and is not altered in nuance or implication by other content of the advertisement.

□   Challenge, and test the impact, of any humor, hyperbole, or emo-tional appeals incorporated in the advertisement—any of which may be interpreted differently than the advertiser intends.

□   Add disclaimers as required, following the four P's of Prominence, Presentation, Placement, and Proximity. Make sure the disclaimer is not adding information critical to understanding the meaning of the claim or information that alters the predictable interpretation of the claim.

□   Timing of the advertising program may take its normal course, unless the competitor changes its product formulation. If the claim is no lon-ger valid, a competitor will no doubt let you know. However, it may be prudent to monitor the truth of the assumptions about the compe-titor's product that underlie the claim, at scheduled dates throughout the planned life of the claim, so as not to leave the business scrambling to respond to an allegation of a false and misleading campaign.

AN OUNCE OF PREVENTION: TROUBLESHOOT YOUR CLAIM BEFORE LAUNCH

Keeping their eye on the upside benefits of comparative advertising, mar-keters may be inclined to let the lawyers worry about the risks. That would be a mistake. Every professional in the company who is part of a comparative advertisement development process has an opportunity to mitigate risk for the benefit of the company and its shareholders.

- ☐ Know the standards of laws and regulations for fair and truthful advertising in your country or local jurisdiction. Look for case studies that may provide guidance for your company's own ambitions.
- ☐ Assess your company's tolerance for taking risks and ability to withstand challenges.
- ☐ Form a realistic and prudent view of the required investment to take your claim to public advertising, including resources for defending the claim if it is later challenged. Calculate costs and rewards of mounting a comparative advertising claim and ensure that the financial analysis can support a business case for proceeding.
- ☐ Examine research-on-hand to generate comparative statements that have the best chance of being supported.
- ☐ Prepare objective defensible evidence to support the claim (or abandon plans if support is not forthcoming).
- ☐ Disaster-check the advertisement for unintended messages that consumers may infer.
- ☐ Negotiate with competitors who complain. Allow a complaint to proceed to litigation only if the company is ready for hostile scrutiny of sensitive information.
- ☐ Have an exit strategy as part of your overall contingency planning, well before the advertisement is launched.
- ☐ Use the resources available online via the government websites, NAD, and get the ASTM, AMA, and the Television Networks claims guidelines as resources to educate yourself about claims substantiation research and processes. Go to conferences and workshops where claims substantiation is the topic.

INTO THE FRAY: PLAYING DEFENSE

Every comparative advertiser should be ready to be challenged. Potential challengers include competitors, government enforcement agencies, or large groups of consumers and the class action attorneys who represent them. Advertising that is particularly aggressive or disparaging of another company or product practically goads adverse scrutiny. In the era of rapid electronic communication, it takes only a few minutes for one of your competitor's executives, agency staff, or employees who see the campaign to fire off a copy to the corporate lawyers. In rapid succession may come a legal demand letter, an NAD or network challenge, or even a lawsuit.

The consequences of a challenge can be quite dramatic, and also expensive. Television networks may pull an advertisement if they determine it contains an advertising claim that is not adequately substantiated. Additionally, a court can issue an injunction order for immediate removal if it determines that the advertisement is false and damaging to a competitor or is likely to be found so at a future trial. Investigations and lawsuits initiated by consumer protection agencies can give rise to significant financial liability either by way of a false advertising verdict or a settlement. Preparation to launch a comparative campaign should incorporate steps of preparation to defend, should a complaint later ensue.

☐ Evaluate at the outset whether the claim is one that is likely to be challenged by a competitor. Is the competitor litigious? Is the claim going to tout superiority as to a product attribute where the competitor claims to have an advantage? If the claim is likely to be controversial, plan for the worst.

☐ Make sure at the outset that there is qualified person (other than counsel) who can explain and defend the testing procedures if necessary.

☐ Make sure you have a reasonable and fair rationale for any key decisions that you make with respect to your test methodology—particularly if they deviate from standard practice.

☐ Make sure that your claim substantiation evidence addresses not only the express statements made in the advertising but also any claims that are reasonably implied.

☐ Be careful in what you say in writing. If you have concerns, pick up the phone to communicate them to other relevant members of your team.

INTO THE FRAY: PLAYING OFFENSE

Whether or not a corporation is inclined to comparative advertising itself, alert attention to competitive activity is an essential component of risk management.

☐ Put into place a corporate watch program, with specific attention to the consistency between words and visuals of advertisements in your corporate sector.

☐ On discovering an offending comparative advertisement, start by direct business communications to the advertiser.

☐ If the offending advertisement is televised, consider the pros and cons of an efficiently implemented complaint to the network.

☐ If more efficient options do not work, assess the trade-offs involved in an action through the regulator or courts. The first decision is whether the business is ready for a public confrontation. Budget required should be prudently estimated; action is frequently more costly than people expect.

VIVE LA DIFFERENCE—ADAPTING COMPARATIVE ADVERTISING TO DIFFERENT COUNTRIES

☐ Famous brands maintain consistent positioning worldwide. The opportunity presents itself for an international company to communicate its desired competitive positioning in different countries. Deciding to take hold of that opportunity is a matter of top-level strategy for the company.

☐ Engage local legal counsel in each country where the campaign is planned to advise on local laws and regulations and to vet copy.

☐ Pretest advertisements in each country for cultural compatibility.

☐ Avoid disparagement anywhere. Notwithstanding inter country differences in nuances of permitted comparisons, almost all countries of interest to international advertisers prohibit disparagement of competitors' products.

SOCIAL MEDIA AND COMPARATIVE ADVERTISING

☐ As a matter of contemporary business strategy, every plan to launch a comparative advertising should consider an accompanying social media component.

☐ Stay alert to the risks of soliciting consumers to participate in online discussions of your product's superiority; such solicitation activities may be subject to regulations affecting comparative advertising through traditional media. Regulations concerning endorsements are also pertinent. The additional challenge for marketers is that there is little control over where consumers take the conversation on social media.

☐ Endorsers that are paid or otherwise associated by business interest to the advertiser must be disclosed.

☐ Administrators of advertising laws and regulations are continuing to be presented with novel cases that, in turn, are shaping the evolving law. Legal counsel should be consulted every time a new creative marketing idea is proposed for social media.

CHAPTER 13

Twenty-First Century Resources

This final chapter compiles a curated bibliography of resources published in the 21st century, from articles to videos to media reports. The set of resources offers a starting point for further research, and a compendium for experts to consult should they be called upon for adjudicating opinions.

DEFINITIONS OF COMPARATIVE ADVERTISING

Defining comparative advertising is a useful starting point for professional training of those responsible for designing advertising or defending it. The significance of definitions in dispute resolution arises when parties disagree about whether an advertisement is "comparative" in calling out a competitor or not comparative. For example, a party may defend its use of comparative words like "better" or "faster" as referring to its own improving line of products or services; a competitor, for its part, may see an implied comparison to its own company's products or services. Explicit definitions of comparative advertising are contained in many applicable laws and regulations. Definitions external to laws or regulations are sometimes referred to in dispute resolution as being indicative of industry understanding. Examples of definitions of comparative advertising that have been referred to in the context of documented decisions include the following:

- Doyle C. *A dictionary of marketing*. 3rd ed. Oxford: Oxford University Press; 2011.
- Financial Times lexicon at: http://lexicon.ft.com/Term?term = comparative-advertising.
- Advertising Standards Canada. *Guidelines for the use of comparative advertising*. Available at: https://www.adstandards.com/en/ASCLibrary/guidelinesCompAdvertising-en.pdf, p. 1.

Ruth M. Corbin, (Editor): Practical Guide to Comparative Advertising
Twenty-First Century Resources, Rebecca Bleibaum and Dr. Ruth Corbin, Principal authors.
DOI: https://doi.org/10.1016/B978-0-12-805471-0.00013-3.

- World Intellectual Property Organization. *Intellectual property issues in advertising.* Available at: http://www.wipo.int/sme/en/documents/ ip_advertising_fulltext.html#advert.

VIDEO AND RADIO BROADCAST EXAMPLES OF COMPARATIVE ADVERTISING IN ACTION

YouTube links to thousands of examples of past campaigns although ones that have been enjoined by courts or regulators are usually taken down. Readers may find these examples helpful:
- "The Age of Persuasion", an advertising expert's advice on finding competitive advantage in parity products: http://www.cbc.ca/ageof- persuasion/episode/season-5/2011/04/15/season-five-all-things-being- equal-the-fascinating-world-of-parity-products-1/.
- Microsoft's The Bing Challenge, at: http://tiny.cc/z10xpy, accessed June 2016.
- Complan competitive advertising in the original Tamil language, at: http://tiny.cc/f20xpy.
- Wendy's US video advertisement claiming superior taste to McDonald's French fries, at: http://tiny.cc/h20xpy.

INTERNATIONAL RESOURCES ON REGULATORY GUIDELINES

Companies who do business in more than one country will find it necessary to check what is specifically permissible in each jurisdiction. There is a general prohibition against "misleading" or "disparaging," in all countries that allow comparative advertising at all, but interpretations of what those words mean or imply can differ across jurisdictions.

Public Authorities

Australian Competition and Consumer Commission, at: https://www. accc.gov.au/consumers/misleading-claims-advertising and https://www. accc.gov.au/business/advertising-promoting-your-business/false-or-mis- leading-statements.
- Russia, expert summary at: http://tiny.cc/p40xpy.
- Diamond S. *Reference manual on scientific evidence.* Washington, DC: The National Academics Press; 2011.

Industry Self-Regulatory Organizations (SROs) in Selected Countries

Country	SROs	Website, as on February 2018
Australia	Advertising Standards Bureau	https://adstandards.com.au/
Belgium	Jury for Ethical Practices on Advertising	https://www.jep.be/nl
Brazil	Conselho Nacional de Autorregulamentaçao Publicitaria	http://www.conar.org.br/
Canada	Advertising Standards Canada	https://www.jep.be/nl
Chile	Consejo de Autorregulación y Ética Publicitaria	http://www.conar.cl/
Colombia	Comisión Nacional de Autorregulación Publicitaria	https://www.ucepcol.com/conarp-cjce
France	Autorité de régulation professionelle de la publicité	https://www.arpp.org/
India	Advertising Standards Council of India	http://tiny.cc/b50xpy
Italy	Istituto dell'Autodisciplina Pubblicitaria	http://www.iap.it/
Mexico	Council of Self-Regulation and Advertising Ethics CONAR AC	http://www.conar.org.mx/que_es_conar
New Zealand	Advertising Standards Authority	http://www.asa.co.nz/
United Kingdom	Advertising Standards Authority (ASA) and Committees of Advertising Practice (CAP)	https://www.asa.org.uk/news/a-quick-guide-to-comparative-advertising.html#.VVDGXvCy7-I, with specific codes at https://www.asa.org.uk/codes-and-rulings/advertising-codes.html
United States	Advertising Self-Regulatory Council	http://tiny.cc/h80xpy

North American Television Networks

- ABC. http://abcallaccess.com/app/uploads/2016/01/2014-Advertising-Guidelines-.pdf.
- CBC Radio Canada. http://www.cbc.radio-canada.ca/en/reporting-to-canadians/acts-and-policies/programming/advertising-standards/.
- ESPN. http://3pam8anagme464n1l3p4ufl4.wpengine.netdna-cdn.com/wp-content/uploads/sites/3/2017/05/2017-ESPN-Advertising-Standards.pdf.
- NBC. https://www.nbcuadstandards.com/guidelines.nbc.

Chamber of Commerce Members Worldwide

- Consolidated ICC Code of Advertising and Marketing Communications Practice at: http://www.codescentre.com/downloads.aspx; 2011.

CASES ADJUDICATED BY COURTS OF LAW AND ADMINISTRATIVE TRIBUNALS

European Court of Justice

- *L'Oréal SA and others v. Bellure NV and others* [2010] EWCA CEV 535, involving parity claims by self-admitting brand-imitator companies.
- ECJ, 8 February 2017, Case C-562/15, *Carrefour Hypermarchés SAS v. ITM Alimentaire International SASU*, a case regarding price comparisons made between stores of different formats, with insufficient disclosure to the advertising audience; illustrates the requirement to compare "like with like."

Australia

- *Australian Competition and Consumer Commission v A Whistle & Co Pty Limited* [2015] FCA 1447, regarding fake testimonials.
- *Gillette Australia Pty Ltd v. Energizer Australia Pty Ltd* [2002] FCAFC 223 (a dispute between two famous battery brands, with respect to alleged misleading by omission of consumer-relevant information).

Canada

- *Commissioner of Competition v. Bell Canada, Bell Mobility Inc. and Bell Expressvu Limited Partnership (Consent Agreement)* (June 28, 2011), CT-2011-005, at http://www.ct-tc.gc.ca/CMFiles/CT-2011-005_Consent%20Agreement_1_45_628-2011_7559.pdf (re insufficiency of disclaimers).

- *Richard v. Time Inc.*, 2012 SCC 8, [2012] 1 S.C.R. 265 (re the "average consumer" test).
- *Yazdanfar v. College of Physicians and Surgeons of Ontario*, [2013] O.J. No. 4787 (Div. Ct.) (re implied comparisons through superlatives as used by medical professionals).

India

- *Pepsi Co., Inc. And Ors. vs Hindustan Coca Cola Ltd. And Anr. on 1 September, 2003, re Pepsi v Pappi*: 2003 (27) PTC 305 Del (re disparagement by implied use of competitor's name).

The United Kingdom

- UK's Sodastream decision, deviating from decisions in other countries: http://www.campaignlive.co.uk/article/clearcast-forces-sodastream-pull-denigrating-ad/1161191#p36k6PEdqY7i2LmS.99; Clearcast's press release announcing decision and appeal at: http://tiny.cc/6n1xpy.

The United States

- *Schick Mfg, Inc. v. Gillette Co.*, 372 F Supp. 2d 273, May 31, 2005 (re claims of superior shaving performance).
- *Church & Dwight Co. Inc. v. Clorox Co.*, case number 11-cv-00092, S.D.N.Y. (2012) (re removal of odors in Super Scoop cat litter).
- *Grubbs v. Sheakley Group, Inc.*, 807 F.3d 785, 798 (6th Cir. 2015); *Pernod Ricard USA, LLC v. Bacardi U.S.A., Inc.*, 653 F.3d 241, 248 (3d Cir. 2011) (re-establishment of harm).

REGULATORY ADJUDICATIONS AND SECONDARY REPORTS ON REGULATORY ADJUDICATIONS
From ASA in Britain

- *Bulldog Communications Ltd.*, March 2, 2005, regarding its claim of "the ultimate broadband experience", referenced at: https://www.asa.org.uk/advice-online/types-of-claims-subjective-or-objective-superlative.html.
- *Kevin Nash Group Plc*, October 27, 2004, regarding its claim of being "the world's leading carp company", referenced at: https://www.asa.org.uk/advice-online/types-of-claims-subjective-or-objective-superlative.html.

- *Wren Living Ltd t/a Wren Kitchens*, July 26, 2017; competitor challenged claim by Wren Kitchens that it was "The UK's Number 1 Kitchen Retail Specialist"; ruling at: https://www.asa.org.uk/rulings/wren-living-ltd-a17-382551.html.
- *Inspop.com Ltd t/a Confused.com*, July 5, 2017; ads through different media stated it was "No.1 for car savings"; challenged for being ambiguous and unverified; analysis published online at: https://www.asa.org.uk/rulings/inspop-com-ltd-a17378098.html.
- *Eden Farmed Animal Sanctuary t/a Go Vegan World*, July 26, 2017; a newspaper advertisement for Go Vegan World, a vegan campaign group featured a photo of a cow behind a piece of barbed wire. The headline text stated "Humane milk is a myth. Don't buy it." Smaller text stated "I went vegan the day I visited a dairy. The mothers, still bloody from birth, searched and called frantically for their babies. Their daughters, fresh from their mothers' wombs but separated from them, trembled and cried piteously, drinking milk from rubber teats on the wall instead of their mothers' nurturing bodies. All because humans take their milk. Their sons are slaughtered for their flesh and they themselves are slaughtered at 6 years. Their natural lifespan is 25 years. I could no longer participate in that. Can you?". Complainants challenged whether the claim "Humane milk is a myth" and the other specific claims were misleading and could be substantiated; ruling published at: https://www.asa.org.uk/rulings/eden-farmed-animal-sanctuary-a17381845.html.

From NAD in the United States

- On Ragu/Prego comparative advertising: http://tiny.cc/vo1xpy.
- On Mom Brands' claim of comparative superiority over other cereals: http://tiny.cc/cp1xpy.
- Safety 1st (Dorel Juvenile Group, Inc.), Case Report #4223, NAD/CARU Case Reports (2004).
- Frito Lay (Lay's Stax), Case Report # 4270, NAD/CARU Case Report (2004), on Frito-Lay's claim of taste superiority over Pringles.
- NAD/CARU's archive of case reports archive, at: http://tiny.cc/9p1xpygi.

From New Zealand's Advertising Standards Authority

- Advertising Standards Authority of New Zealand, Complaint 11/475, AWAP 1108, *Foodstuffs (Auckland) Ltd. v. Progressive Enterprises*; price comparisons found to be ambiguous and misleading; this decision and others searchable at: http://old.asa.co.nz/search_code.php.

- Advertising Standards Authority of New Zealand, Complaint 12/228, K. Fry and Others v. Guardians NZ Limited; the advertiser was a dog minding service comparing itself to dog kennels; its advertisement was found to mislead consumers through fear-mongering about the risks of boarding dogs at kennels; this decision and others searchable at: http://old.asa.co.nz/search_code.php.

MEDIA REPORTS AND PRESS RELEASES ON CAMPAIGNS AND CASES

- AdAge, *"Court Rules against Gillette Razor Package Claim"*, June 23, 2005, re follow-up to early enjoinment of "superior shaving" advertisement; http://adage.com/article/news/court-rules-gillette-razor-package-claim/46118/.
- Associated Press report on the "hot dog wars," August 15, 2011, communicated by CBC News/Business at: http://tiny.cc/uq1xpy, accessed November 27, 2012.
- Ruth M. Corbin, on context effects in people's judgments: Context Effects on Validity of Response: Lessons from focus groups and complacent frogs. *Vue*; November 2006, p. 10−14; accessible online at: http://corbinpartners.com/wp-content/uploads/2012/12/vue1106.pdf.
- Federal Trade Commission press release, "Reebok to pay $25 Million in Customer Refunds to Settle FTC Charges of Deceptive Advertising;" September 28, 2011, at: http://tiny.cc/dr1xpy, re Reebok's claim of sneakers that tone muscles.
- Financial Post, on Bell Canada's record ten-million dollar fine: http://tiny.cc/sr1xpy.
- Murdoch S. National Australia Bank set to sign up its one millionth customer since the 'break-up' campaign. *The Australian*; July 24, 2010.
- New York Times, on Dove Body Wash comparative campaign implicating competitor Dial: June 25, 2013, at: http://tiny.cc/as1xpy [last accessed December 2017].
- Vaas S. on Scottishjobs.com's claim of being "Number 1 for Plum Jobs": "Jobs Agency Falsified its CV." *The Sunday Herald*; August 8, 2004.
- Jackie Wattles on Luminosity's unsupported claims of improvement in cognitive function and staving off Alzheimer's, CNN Tech report available at: http://tiny.cc/kt1xpy.

ACADEMIC JOURNAL REPORTS: HISTORY AND EFFECTIVENESS OF COMPARATIVE ADVERTISING

- Barry TE. Comparative advertising: what have we learned in two decades? *J Adv Res* March–April 1993;**33**(2):19–29.
- Beard F. Practitioner views of comparative advertising. *J Adv Res* 2013;**53**(3):313–23.
- Beard F. A history of comparative advertising in the United States. *Journ Commun Monogr* 2013;**15**(3):114–216.
- Kalro AD, Sivakumaran B, Marathe RR. The ad format-strategy effect on comparative advertising effectiveness. *Eur J Market* 2017;**51**(1):99–122.
- Shafiulla B. Comparative advertising: an analysis of cases of disparagement. *IUP J Market Manage* May 2013;**12**(2), on the risk of actionable disparagement.
- Williams K, Page RA. Comparative advertising as a competitive tool. *J Market Dev Compet* 2013;**7**(4).

ACADEMIC JOURNAL REPORTS: EXPERT EVIDENCE AND LEGAL PRECEDENTS

- Ennis JM, Ennis DM. Justifying count-based comparisons. *J Sens Stud* 2017;**27**:130–6.
- Mishra H, Corbin R. Online surveys in Intellectual Property Litigation: Doveryai No Proveryai. *Trademark Reporter* October 2017;**107**; the article assembles current practices worldwide regarding reliability and validity of Internet surveys.
- Leighton RJ. Materiality and puffing in Lanham Act False Advertising Cases: the proofs, presumptions and pretexts. *Trademark Reporter* 2004;**94**(3):585–633.
- Passman N. Supporting advertising superiority claims with taste tests. *Food Technol* August 1994:71–74.

ACADEMIC JOURNAL REPORTS: CROSS-CULTURAL DIFFERENCES

- Singh M. Comparative advertising effectiveness with legal and cross-culture framework. *Int J Res Manage Pharm* 2014;**3**(3).
- Jeon JO, Beatty SE. Comparative advertising effectiveness in different national cultures. *J Business Res* 2002;**55**:907.

SLIDE PRESENTATIONS AND PUBLIC ADDRESSES

- Choudhary N. http://tiny.cc/2t1xpy, last visited December 2017 (prepared by an Indian student researcher).
- Fair LA. "FTC Update: Enforcement Priorities and Key Cases" Speaker at National Advertising Division Annual Conference, "What's New in Advertising Law, Claim Support and Self-Regulation?" New York City; September 29–30, 2014.
- Musgrove J, Edmondstone D. "The Shifting General Impression of Disclaimers," address to the Canadian Bar Association 2012 Competition Law Spring Forum: Best Practices in a Time of Active Enforcement; May 2, 2012.
- Pritchard B, Corbin R. "Avoiding Landmines. How to Support Comparative Advertising Claims," presentation to a conference sponsored by Advertising Standards Canada; February 27, 2012. Content available online in slide form at: http://corbinpartners.com/wp-content/uploads/2012/12/ASC-Presentation-Avoiding-Landmines-How-to-Support-Comparative-Advertising-Claims.pdf.
- Civil T (Frito-Lay Canada), Hearn B (McMillan Binch LLP). "The Universe's Best Guide to Claim Substantiation Strategies", Address to the Canadian Institute' 11th Annual Advertising and Marketing Law Conference, Toronto; January 25–26, 2005.
- Reed W. "Bullet-Proofing Your Ad Substantiation and Performance Claims," Address to the Canadian Institute's 10th Annual Advertising and Marketing Law Conference; January 22–23, 2004.

TEXTS ON ADVERTISING LAW

- Pritchard B, Vogt S. *Advertising and marketing law in Canada.* Markham: LexisNexis Canada, Inc.; 2004.

TEXTS ON RESEARCH DESIGN PERTINENT TO EXPERT EVIDENCE

- Beecher-Monas E. *Evaluating scientific evidence: an interdisciplinary framework for intellectual due process.* New York: Cambridge University Press: 2007.
- Corbin R, Gill AK. *Survey evidence and the law worldwide.* Butterworth; 2008.

- Corbin R. Surveys and other marketplace evidence. In: Cameron D, editor. *Canadian trademark benchbook*. Toronto, ON: Carswell; 2014.
- Finkelstein MO. *Basic concepts of probability and statistics in the law.* New York: Springer; 2009. Kadane JB. *Statistics in the law.* New York: Oxford University Press; 2008.
- Meilgaard M, Civille GV, Carr BT. *Sensory evaluation techniques.* 5th ed. Boca Raton, FL: CRC Press; 2015.
- Lawless H, Heymann H. *Sensory evaluation of food: principles and practices.* 2nd ed. New York: Springer Press; 2010.
- Stone H, Bleibaum R, Thomas H. *Sensory evaluation practices.* 4th ed. Cambridge: Elsevier/Academic Press; 2012.

BUSINESS-BASED AUTHORITIES AND RESOURCES ON STANDARDS

Text in *italics* is taken from the websites of organizations serving the business community, as on August 2018.

ASTM International (ASTM)

ASTM was first organized in 1898 and has grown into one of the largest voluntary standards development systems in the world. It is a nonprofit organization that provides a forum for producers, users, ultimate consumers and those having a general interest (representatives of government and academia) to meet on common ground and write standards for materials, products, systems, and services, and the promotion of related knowledge.

ASTM International believes that technically competent standards result when a full consensus of all concerned parties is achieved and rigorous due process procedures are followed. This philosophy and standards development system ensure technically competent standards have the highest credibility when critically examined and used as the basis for commercial, legal, or regulatory actions. ASTM International standards are developed and used voluntarily. Standards become legally binding only when a government body references them in regulations or when they are cited in a contract.

ASTM International publishes more than 12,000 standards under over 130 standards-writing committees each year. ASTM International standards are subject to revision at anytime by the responsible technical committee and must be reviewed every five years, and if not revised, either reapproved or withdrawn.

ASTM Committee, E18 on Sensory Evaluation

Committee E18 is a technical committee of ASTM International, designed to promote knowledge, stimulate research, and develop principles and standards for the sensory evaluation of materials and products. Committee E18 is comprised of nearly 250 industry and academia professionals-food scientists, sensory scientists, psychophysicists, statisticians, psychologists, and other professionals, representing the world's leading universities and Fortune 500 companies. These professionals are at the forefront of new product development technology, designing and applying the appropriate sensory methods for consumer goods including food, beverage, tobacco, household and personal care products, among others, worldwide.

ASTM Document, E1958 Standard Guide for Sensory Claims Substantiation

E1958 was developed and approved by the collective membership of ASTM Committee E18, individuals who are intimately involved with the design and analysis of studies to assess product performance, and who are responsible for the interpretation and communication of their research results to the business and professional communities.

In the spring of 1993, Committee E18 held a discussion on the increased interest in sensory testing to support advertising claims. Although a number of individuals and groups had made recommendations on how to effectively conduct sensory tests for advertisement claims, there were many inconsistencies between groups. Committee E18 is composed of sensory professionals whose purpose is to write voluntary industry standards for this field, it seemed logical that they should attempt to review, combine, and filter individual and group recommendations into one document. Those contributing to this document represent both large and small corporations, academicians, statisticians, attorneys, and consultants in a wide variety of consumer products categories. The categories include but are not limited to food, beverage, cosmetics, health and beauty aids, and other related products as well as sensory and consumer research methods used to substantiate claims research.

ESOMAR

ESOMAR is the world association for market, social, and opinion researchers. Founded in 1948, ESOMAR began as a regional association within Europe. Hence, the name ESOMAR was originally an acronym for European Society for Opinion and Market Research. Currently, with more than 4900 members in over 130 countries, ESOMAR's global membership brings together professionals in market and opinion research, marketing, advertising, business, public affairs, and media from across the world.

ESOMAR provides ethical guidance and actively promotes self-regulation in partnership with a number of associations across the globe. All ESOMAR members agree to abide by the ICC/ESOMAR International Code on Market and Social Research, which has been jointly drafted by ESOMAR and the International Chamber of Commerce and is endorsed by the major national and international professional bodies around the world.

The ICC/ESOMAR Code on Market and Social Research, which was developed jointly with the International Chamber of Commerce, sets out global guidelines for self-regulation for researchers and has been undersigned by all ESOMAR members and adopted or endorsed by more than 60 national market research associations worldwide.

In addition to the ICC/ESOMAR Code, the ESOMAR Professional Standards team also produces and promotes industry guidelines to assist researchers in addressing the legal, ethical, and practical considerations in fast developing areas such as online, mobile and social media research. The Professional Standards team also manages the disciplinary procedures regarding complaints about members.

International Standards Organization (ISO)

ISO is an independent, non-governmental international organization with a membership of 161 national standards bodies. Through its members, it brings together experts to share knowledge and develop voluntary, consensus-based, market relevant International Standards that support innovation and provide solutions to global challenges.

As of this writing, the ISO is undertaking to set guidelines for sensory and consumer claims that may complement the ASTM E1958 standards. ISO intends a multinational standard that will enhance alertness to the substantiation needed for advertised claims of sensory experience.

GUIDELINES FROM INTERNET SEARCH ENGINES, SOCIAL MEDIA SPONSORS, AND MAJOR ONLINE VENDORS

Google

Until 2016 Google's advertising policy (http://archive.is/YzUpG; last accessed August 2018), explicitly excluded "promotions that contain the superlatives and comparatives 'best,' '#1,' 'better than,' 'faster than,' or any other equivalent claims where that claim is not supported by third-party verification on the landing page." By way of example given by Google, advertising text reading "#1 window cleaning service in the world" with no evidence on the landing page to support the claim of being #1 would only be permissible by Google if the landing page linked to a third-party industry analysis that showed the company in question to be the most popular, highest quality service, etc. Google has since simplified the wording of its policy to prohibit any advertising that misrepresents or misleads (https://support.google.com/adwordspolicy/answer/6020955?hl= en&topic=1310871&ctx=topic&visit_id=1-636552831927079793-2462 498657&rd=1; last accessed February 2018).

Bing

Bing/Microsoft's policy on comparative advertising is contained within its discussion of use of others' trademarks on the Internet. "Microsoft allows the fair use of trademarks in ad text, such as.... comparative advertising, when supported by independent research" (https://advertise.bingads. microsoft.com/en-ca/resources/policies/intellectual-property-policies; last accessed February 2018).

Yahoo

Like Bing, Yahoo puts a focus on the use of a competitor's trademark for comparison purposes, allowable "so long as the comparison is fair, accurate, and supported by independent research." Product descriptions must also be "truthful and accurate" in the context of any comparisons being made (https://guce.oath.com/collectConsent?brandType=eu&. done=https%3A%2F%2Fau.adspecs.yahoo.com%2F%3Frnd%3D1%26guc-counter%3D1&sessionId=2_cc-session_54974938-8892-47c6-a074-ffadb-b6860e2&lang=en-US&inline=false; last accessed August 2018).

Facebook

Facebook prohibits advertisements (https://www.facebook.com/policies/ ads/) that contain "content that infringes upon or violates the rights of any third party, including copyright [and] trademark" rights. Advertisements are prohibited if they have "deceptive, false, or misleading content, including deceptive claims, offers, or business practices." Furthermore, "ads, landing pages and business practices must not contain deceptive, false or misleading content," nor "exaggerated claims." Comparative claims are not called specifically.

LinkedIn

LinkedIn prohibits ads which are "fraudulent, deceptive or include misleading titles, statements or illustrations. Your product or service must accurately match the text of your ad—don't lie, don't exaggerate and don't make false claims. The photos and images in your ad should have a reasonable relationship to the product or service being advertised. The claims you make in your ad should have factual support. Do not make deceptive or inaccurate claims about competitive products or services" (https://www.linkedin.com/legal/sas-guidelines).

Twitter

Twitter permits the use of competitors' trademarks for comparative advertising only when supported by independent research (https://business.twitter.com/en/help/ads-policies/other-policy-requirements/promoted-trends-guidelines.html).

Amazon

All information contained with the creative content of the advertisement "must be accurate and verifiable. Performance claims and comparisons also require substantiation on the creative or the landing page. Objective superlative claims always require the source and date of the evidence to be cited... Placing the citation on the ad unit is preferred as it provides the best customer experience but, where ad space is limited, placing the citation on the landing page instead is acceptable. The evidence must be sufficiently recent to support the claim" (https://advertising.amazon.com/ad-specs/en/policy/creative-acceptance).

INDEX

Printed in the United States
By Bookmasters